HUMAN MOLECULAR BIOLOGY LABORATORY MANUAL

Human Molecular Biology Laboratory Manual

Stefan Surzycki

Department of Biology
Indiana University

Blackwell
Publishing

© 2003 by Blackwell Science Ltd
a Blackwell Publishing company

350 Main Street, Malden, MA 02148-5018, USA
108 Cowley Road, Oxford OX4 1JF, UK
550 Swanston Street, Carlton South, Melbourne, Victoria 3053, Australia
Kurfürstendamm 57, 10707 Berlin, Germany

The right of Stefan Surzycki to be identified as the Author of this Work has been asserted in accordance with the UK Copyright, Designs and Patents Act 1988.

All rights reserved. No part of this publication may be reproduced, stored in a retrieval system, or transmitted, in any form or by any means, electronic, mechanical, photocopying, recording or otherwise, except as permitted by the UK Copyright, Designs and Patents Act 1988, without the prior permission of the publisher.

First published 2003 by Blackwell Science Ltd

Library of Congress Cataloging-in-Publication Data
Surzycki, Stefan, 1936–
    Human molecular biology laboratory / Stefan Surzycki.
      p. cm.
    Includes bibliographical references and index.
    ISBN 0-632-04676-7 (alk. paper)
    1. Molecular biology – Laboratory manuals.   2. Human genetics – Laboratory manuals.   I. Title.
QH506 .S893 2002
611'.01816 – dc21

                                                                               2002071215

ISBN 0-632-04676-7

A catalogue record for this title is available from the British Library.

For further information on
Blackwell Publishing, visit our website:
www.blackwellpublishing.com

Contents

Preface	xi
Chapter 1 Preparation of Human Genomic DNA	1
Introduction	1
Background	1
Cell Breakage	3
Removal of Protein	3
Deproteinization using organic solvents	3
Deproteinization using enzymes	6
Removal of RNA	6
Concentrating the DNA	7
Determination of the Purity and Quantity of DNA	9
First Laboratory Period	11
Safety precautions	11
Technical tips	11
Protocol	13
Second Laboratory Period	16
Safety precautions	16
Technical tips	16
Protocol	16
Expected results	18
References	19
Chapter 2 DNA Fingerprinting: Multi-locus Analysis	20
Introduction	20
Background	20
First Laboratory Period	25
Experiment 1: restriction enzyme digestion	25
Introduction	25
Background	25
Technical tips	26
Protocol	27
References	28

Experiment 2: agarose gel electrophoresis	29
Introduction	29
Background	29
Safety precautions	34
Protocol	34
References	36
Second Laboratory Period	37
Experiment 3: Southern blotting	37
Introduction	37
Background	37
Safety precautions	38
Technical tips	38
Protocol	38
References	41
Third Laboratory Period	43
Experiment 4: preparation of probe and hybridization	43
Introduction	43
Background	43
Technical tips	50
Protocol	51
References	54
Fourth Laboratory Period	56
Protocol	56
Fifth Laboratory Period	58
Protocol	58
Data analysis	59

Chapter 3 DNA Fingerprinting: Single-locus Analysis   62

Introduction	62
Background	63
First Laboratory Period	66
Protocol	66
Second Laboratory Period	69
Protocol	69
Data analysis	69
Expected results	72
References	75
General reading	75

Chapter 4 Out of Africa: Origin of Modern Humans   76

Introduction	76
Background	76
Origin of humans	76
PCR	80

First Laboratory Period	85
Technical tips	85
Protocol	85
Second Laboratory Period	89
Safety precautions	89
Protocol	89
Data analysis	92
Expected results	93
References	94

Chapter 5 DNA Sequencing — 95

Introduction	95
Background	96
DNA sequencing methods	97
Sequencing strategies	98
References	99
First Laboratory Period	100
Experiment 1: nebulization shearing of DNA	100
Introduction	100
Background	100
Safety precautions	102
Technical tips	103
Protocol	104
Expected results	106
Experiment 2: repair of the ends of sheared DNA	106
Introduction	106
Safety precautions	106
Technical tips	107
Protocol	108
References	110
Second Laboratory Period	111
Experiment 3: ligation to sequencing vector	111
Introduction	111
Background	111
Technical tips	118
Protocol	119
References	121
Experiment 4: transformation of bacteria by electroporation	121
Introduction	121
Background	122
Technical tips	124
Protocol	126
Expected results	128
References	129

Third Laboratory Period		130
Experiment 5: preparation of plasmid for DNA sequencing		130
Introduction		130
Background		130
Technical tips		131
Protocol		132
Experiment 6: sequencing reactions for an ABI 3700 sequencer		134
Introduction		134
Technical tips		135
Protocol		136
Fourth Laboratory Period		137
Experiment 7: removing dideoxy terminators		137
Protocol		137
References		137

Chapter 6 Computer Analysis of Sequencing Data — 139

- Introduction — 139
- Background — 139
  - Databases and sequence formats — 140
  - Sequence alignments — 143
  - BLAST — 147
  - FASTA — 150
  - BLAST versus FASTA — 150
  - Single-sequence analysis — 151
  - Dot matrix analysis — 152
  - Technical tips — 154
  - Protocol — 155
  - References — 157
- Sequence Alignment with BLAST — 158
  - Search of "nr" database — 158
  - Search for an *Alu* SINE element — 159
  - Search for expressed sequences — 159
- Search for the Chromosome Position of the Query Sequence — 159
- Single-sequence Analysis — 161
  - Converting file formats — 161
  - Base content analysis — 162
  - Restriction enzyme site analysis — 162
  - Dot matrix analysis — 162

Chapter 7 Determination of Human Telomere Length — 164

- Introduction — 164
- Background — 165
- First Laboratory Period — 170
  - Experiment 1: isolation of genomic DNA — 170

Introduction	170
Background	170
Safety precautions	171
Protocol	171
Second Laboratory Period	174
Experiment 2: determination of DNA concentration and purity	174
Protocol	174
Experiment 3: restriction enzyme digestion	174
Introduction	174
Background	175
Technical tips	175
Protocol	176
Experiment 4: agarose gel electrophoresis	177
Introduction	177
Background	177
Safety precautions	177
Technical tips	177
Protocol	178
Third Laboratory Period	180
Experiment 5: Southern transfer	180
Introduction	180
Background	180
Safety precautions	180
Technical tips	181
Protocol	181
Fourth Laboratory Period	185
Experiment 6: DNA hybridization	185
Introduction	185
Background	185
Technical tips	185
Protocol	186
Fifth Laboratory Period	187
Protocol	187
Sixth Laboratory Period	190
Experiment 7: analysis of TRF length	190
Introduction	190
Background	190
Technical tips	191
Protocol	192
References	194

Chapter 8 RT-PCR of Human Genes — 196

Introduction — 196
Background — 196

First Laboratory Period		198
Experiment 1: purification of total RNA		198
Introduction		198
Background		198
Safety precautions		200
Technical tips		201
Protocol		202
References		203
Experiment 2: RNA agarose gel electrophoresis		204
Introduction		204
Background		204
Safety precautions		205
Technical tips		205
Protocol		205
References		207
Second Laboratory Period		209
Experiment 3: running an RT-PCR		209
Introduction		209
Background		209
Safety precautions		212
Technical tips		212
Protocol		213
Appendix		215
DNA Purification		215
Equipment and supplies		215
Solutions to prepare		216
DNA Fingerprinting: Multi-locus Analysis		217
Equipment and supplies		217
Solutions to prepare		218
DNA Fingerprinting: Single-locus Analysis		219
Equipment and supplies		219
Solutions to prepare		219
Out of Africa: Origin of Modern Humans		220
Equipment and supplies		220
Solutions to prepare		220
DNA Sequencing		220
Equipment and supplies		220
Solutions to prepare		221
Determination of Human Telomere Length		222
Equipment and supplies		222
RT-PCR of Human Genes		222
Equipment and supplies		222
Index		223

Preface

The recent completion of the human genome-sequencing project is an important development in the history of biological sciences. It will not only promote the understanding of the human genome, but will also profoundly change the discipline of molecular biology and affect medical practices. The human genome is of great interest and is the subject of intensive basic and applied research. The molecular biology techniques used in this research are highly advanced and unique. Learning these techniques will permit students to learn the basic principles of molecular biology and will prepare them to work with the human genome.

These skills are in great demand by biotechnology, forensic laboratories, and pharmaceutical companies. This laboratory manual provides the student with basic experience in and an understanding of cutting-edge techniques in molecular biology. In addition, the experiments described in this manual will provide students with an opportunity for analyzing and studying their own genes.

The goal of this laboratory manual is not only to teach basic molecular biology techniques, but also to convey the excitement of performing experiments and comparing the results to a large body of data collected about the human genome.

The topics of the course revolve around a central theme of analysis of the student's own genome, i.e. its structure and gene expression. These topics include eight exercises.

1. Preparation of genomic DNA. Cheek cells are the source of this DNA. Collecting these cells is a non-invasive procedure that makes it possible to use DNA purification in a classroom situation. The techniques that are used in the course of this experiment are large-scale purification of DNA, spectroscopic analysis of DNA, and determination of DNA concentration and purity.

2. DNA fingerprinting using multi-locus analysis with a human variable number tandem repeat probe. Students use their own DNA for this analysis. In this procedure students learn the techniques of Southern blot transfer, preparation of non-radioactive probes, hybridization, and chemiluminescent autoradiography. Use of a non-radioactive probe removes the difficulties of working with radioactive materials in the class environment. It also eliminates the problem of disposing of a large quantity of radioactive waste that will invariably be generated when working with a large class. Moreover, the non-radioactive procedure is a more advanced technique that has recently been finding general acceptance in basic research and industry.

3. DNA fingerprinting with a single-locus probe. This technique is used in standard forensic analysis. The probe used is the standard forensic D2S44 probe. It represents a tandem repeat region that is present on human chromosome 2. Students will learn methods of forensic profiling and analyze data using a fixed bins database of allele frequencies prepared for this probe by the FBI.

4. Linkage disequilibrium analysis using the DNA markers *Alu* CD4 and the TTTTC repeat. This experiment is based on the paper of Tishkoff et al. (1996) (see the reference section in Chapter 4). The authors introduced this innovative technique in determining a common and recent African origin for all non-African human populations. Analysis of the data consists of the calculation of linkage disequilibrium for the entire class. The results are compared to the disequilibrium found in different world populations. During the course of this experiment, students learn how to perform PCR (polymerase chain reactions), use the thermal cycler, and analyze products using high-resolution agarose gel electrophoresis.

5. Sequencing of human DNA using an ABI capillary sequencer and Big Dye technology. The goal of this experiment is to sequence human DNA using the same procedures employed in large sequencing projects. Students prepare their DNA for sequencing using the random sequencing strategy used by the Human Genome Project. The techniques used in the course of this experiment are preparation of a random sequencing library by nebulization, cloning DNA fragments into a sequencing plasmid, transformation of *Escherichia coli* cells by electroporation, preparation of plasmid DNA for sequencing, and PCR cycle sequencing.

6. Computer analysis of sequencing data. Students carry out local (Basic Local Alignment Search Tool or BLAST) and global alignment analysis using their own sequencing data. They analyze direct and inverted repeats in their DNA by dot plots and learn how to use various databases available for the analysis of human DNA sequences (ESTSDB, ALUDB, ICRDB, etc.).

7. Determination of human telomere length. Telomere length is a reflection of the "mitotic clock" of normal somatic cells and is therefore age dependent. In the course of this experiment, students determine the telomere length of their DNA. The techniques used are multi-enzyme digestion of genomic DNA, turbo-blot transfer, hybridization using an oligonucleotide probe, and computer determination of average telomere length.

8. Analysis of the expression of the β-actin gene in human cheek cells. This determination is carried out using single-tube RT-PCR. Students carry out isolation of total RNA from cheek cells, determine its purity and

concentration, perform RT-PCR reactions, and analyze the results by gel electrophoresis.

The manual is an outgrowth of a semester course taught each year to undergraduate students at Indiana University. Each of the eight experiments constitutes an integrated unit performed in one or more laboratory sessions. The laboratory sessions are designed to meet twice a week for 4 hours and are designed for a limit of 20 students per class. Occasionally students (or instructors) will need to spend additional time in the laboratory in order to finish experiments or to collect results. These times are indicated in the outline for each procedure. The descriptions of the laboratory procedures assume that students will perform all the steps of the procedure. However, at the discretion of the instructor, pre-preparing some materials (e.g. pre-paration of labeled probes, preparation of plasmids for sequencing, etc.) can reduce the session times and session numbers.

In this manual I try to go beyond cookbook recipes for each technique. The description of each technique includes an overview of its general importance, historical background, and theoretical basis for each step. This is done in the hope that students will acquire enough of an understanding of the theoretical mechanisms that they will be able to go on to design their own modifications and methods.

All of the procedures in this book have been used extensively in the teaching of undergraduate laboratories and passed the ultimate test for "working" in the hands of several generations of undergraduates. The descriptions of each step in the protocols are very specific and detailed as to how to carry them out. These instructions may appear to be overly detailed, but they have been developed because of years of experience teaching undergraduates and trying to ensure that the experiments work in inexperienced hands the first time they are performed. In addition, technical tips for carrying out each procedure are incorporated into the text.

In the course preparation I make extensive use of commercially available kits. There are several reasons for their use. First, kits save enormous time in preparation and afford substantial savings in the cost of reagents. Frequently the cost of individual reagents necessary for preparing a laboratory for a large class exceeds the cost of the kit. Second, when using reagents from supply companies, the expertise of their technical support is only a World Wide Web page or telephone call away.

In the manual, I also recommend the use of some instruments for class use. This is generally guided by the usefulness of this instrument in a classroom environment, as well as cost itself.

Laboratory Safety

Anybody using this manual should be familiar with and should follow laboratory safety procedures. Instructors should be familiar with all national, state, local, and university regulations and practices. This is particularly important when disposing of waste (e.g. ethidium bromide, phenol, etc.) and working with human cells. Students should use this manual under instructor supervision. Using some instruments, such as an electrophoresis power supply or high-speed centrifuge, without knowledge of the instrument and proper training or supervision can be very dangerous.

In addition, the description of each experiment includes a section on safety precautions. Before performing any procedure, students should make themselves familiar with its content.

Acknowledgments

Finally, I wish to thank my wife Judy A. Surzycki without whose help and encouragement this book would never have been written. I would also like to express my gratitude to the reviewers of this book, Drs R. Anderson, M. Mehdy, J. Mordacq, J. Normanly, C. Passavant, T. Robson, J.-D. Rochaiz, N. Talbot, and M. Zavanelli, for their helpful comments and suggestions.

CHAPTER 1

Preparation of Human Genomic DNA

Introduction

The goal of this experiment is to isolate and purify a large quantity of high molecular weight human DNA. The source of the DNA will be your cheek cells obtained from a saline mouthwash (a bloodless and non-invasive procedure). You will also learn how to determine DNA concentration and purity. The isolated DNA will be used in several experiments that you will carry out later in the course. The procedure is a "hybrid" between the phenol and chloroform extraction methods, preceded by proteinase K digestion.

This experiment will be performed during two laboratory periods. The first period will include the procedure for cell collection and the initial steps of DNA purification. During the second laboratory period students will finish DNA isolation and determine DNA concentration and purity. Figure 1.1 presents a schematic outline of the experiment.

Background

DNA constitutes a small percent of the cell material and is usually localized in a defined part of the cell. In procaryotic cells DNA is highly condensed and localized in a structure called the nucleoid, which is not separated from the rest of the cell sap by a membrane. In eucaryotic cells the bulk of DNA is localized in the nucleus, which is separated from the rest of the cell sap by a complicated membrane structure. Usually approximately 90 percent of the DNA is localized in the nucleus (chromosomes); the rest can be separated into other organelles such as mitochondria or chloroplasts. In viruses and bacteriophages, DNA is encapsulated by the protein coat and constitutes between 30 and 50 percent of the total mass of the virus. The amount of DNA, as a percent of the total mass of cell material in procaryotes and eucaryotes, is much smaller than that of viruses and is less than 1 percent.

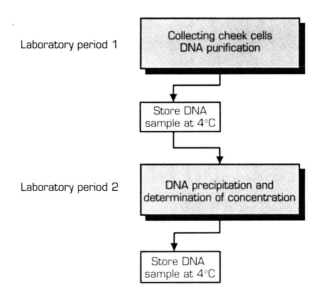

Figure 1.1 Schematic outline of the DNA isolation procedure.

Table 1.1 Composition of living cells

Component	Percent of total cell weight	
	E. coli cells	HeLa (human) cells
Water	70.0	70.0
Inorganic ions	1.0	1.0
Amino acids	0.4	0.4
Nucleotides	0.4	0.4
Lipids	2.2	2.8
Proteins	15.0	22.3
RNA	6.0	1.7
DNA	1.0	0.85

For example, the approximate composition of rapidly dividing *Escherichia coli* cells and human cells (HeLa) is given in Table 1.1.

The goal of DNA purification is to separate DNA from all of the components of the cell listed in Table 1.1. There is no difficulty in separating DNA from small molecules since the molecular weight of DNA is very large. Thus, the components that constitute "major impurities" and should be removed are protein and RNA.

There are several methods of purification of DNA that exploit differences in the physical properties between DNA and proteins. All methods of purification involve five essential steps.

1. Cell breakage.
2. Removal of protein.
3. Removal of RNA.
4. Concentration of DNA.
5. Determination of the purity and quantity of DNA.

Cell Breakage

Cell breakage is one of the most important steps in the purification of DNA. The usual means of cell opening, such as sonication, grinding, blending, or high pressure, cannot be used in DNA purification. These procedures apply strong forces to open cells that shear DNA into small fragments. The best procedure for opening cells and obtaining intact DNA is through application of chemical (detergents) and/or enzymatic procedures. Detergents can solubilize lipids in cell membranes resulting in gentle cell lysis. In addition, detergents have an inhibitory effect on all cellular DNases and can denature proteins, thereby aiding in the removal of proteins from the solution. The lysis of animal cells is usually performed using anionic detergents such as SDS (sodium deodecyl sulfate) or Sarcosyl (sodium deodecyl sarcosinate).

Removal of Protein

The second step in purification involves removing a major contaminant, namely protein, from the cell lysate. This procedure is called deproteinization. Removal of proteins from the DNA solution depends on differences in the physical properties between nucleic acids and proteins. These differences are differences in solubility, differences in partial specific volume, and differences in sensitivity to digestive enzymes.

Deproteinization using organic solvents

The most frequently used methods for removing proteins explore the solubility differences between proteins and nucleic acids in organic solvents. Nucleic acids are predominantly hydrophilic molecules and are easily soluble in water. Proteins, on the other hand, contain many hydrophobic residues making them partially soluble in organic solvents. There are several methods of deproteinization based on this difference and they vary by the choice of the organic solvent.

The organic solvents commonly used are phenol and chloroform containing 1 percent isoamyl alcohol. The method that uses phenol as the deproteinizing agent was introduced by Kirby (1957) and is usually referred to as the **Kirby method**. Use of chloroform isoamyl alcohol mixtures was introduced by Marmur (1961) and is named the **Marmur method**. These methods have undergone many modifications and improvements from the time of their first publications so they bare little resemblance to the original descriptions.

The application of phenol in the Kirby method is based on the following principle. Phenol is crystalline at room temperature, but in the presence of 20 percent water it forms an aqueous suspension containing phenol micelles surrounded by water molecules. Protein molecules generally contain many hydrophobic residues, which are concentrated in the center of the molecule. When an aqueous protein solution is mixed with an equal volume of phenol, some phenol molecules are dissolved in the aqueous phase (approximately 20 percent water and 80 percent phenol). Yet the phenol molecules are extremely hydrophobic. Consequently, they tend to be more soluble in the hydrophobic cores of the protein than in water. As a result, phenol molecules diffuse into the core of the protein causing the protein to swell and eventually to unfold or denature. The denatured protein, with its hydrophobic groups exposed and surrounded by micelles of phenol, is far more soluble in the phenol phase than in the aqueous phase. As a result proteins are partitioned into the phenol phase leaving the nucleic acids in the aqueous phase. Nucleic acids do not have hydrophobic groups at all and are insoluble in the phenol phase.

Application of the phenol method does require mixing the phenol phase with the water phase. This introduces some shearing of DNA molecules. Since only relatively small amounts of protein can dissolve in a given volume of phenol, repeated extraction of the aqueous phase with phenol is required in order to remove all the protein. Because the phenol phase at saturation contains 20 percent water every phenol extraction will remove 20 percent of the DNA into the phenol phase. Even more DNA is lost by entrapment in the interphase layer of precipitated proteins or when the pH of phenol drops below pH 8.0.

Another drawback of the Kirby method is that the oxidation products of phenol can react chemically with DNA (and RNA) molecules. In addition, phenol is highly toxic and requires special disposal procedures.

In order to minimize these effects, several modifications have been introduced.

1. The use of ionic detergents. These detergents, by unfolding the protein, help to expose hydrophobic regions of the polypeptide chains to phenol micelles, thereby aiding partitioning of proteins into the phenol phase.

2. Enzymatic removal of proteins before phenol extraction. This reduces the number of extractions needed, thus limiting the loss and shearing of DNA.

3. Addition of 8HQ (8-hydroxyquinoline) to the phenol. This increases the solubility of phenol in water. In the presence of this compound phenol remains liquefied at room temperature with only 5 percent water. In addition,

8HQ is easily oxidized and, therefore, it plays the role of an anti-oxidant, protecting phenol against oxidation. Since the reduced form of 8HQ is yellow and the oxidized form is colorless, the presence or absence of yellow color is an excellent visual indicator of the oxidation state of phenol.

4. Removal of oxidation products from phenol and prevention of oxidation upon storage or during phenol extraction. Because water-saturated phenol undergoes oxidation rather easily, particularly in the presence of buffers such as Tris, phenol used for DNA purification is twice distilled, equilibrated with water, and stored in the presence of 0.1 percent 8HQ.

5. Adjusting the pH of water-saturated phenol solution to above pH 8 by equilibration of the liquefied phenol with a strong buffer or sodium borate. DNA obtained by the Kirby method is usually of high molecular weight, but contains approximately 0.5 percent protein impurities that can be removed by another method.

The application of a chloroform:isoamyl alcohol (CIA) mixture in the Marmur deproteinization method is based on a characteristic of this organic solvent that differs from phenol. The chloroform is not miscible with water and, therefore, even numerous extractions do not result in DNA loss into the organic phase. The deproteinization action of chloroform is based on the ability of denatured polypeptide chains to partially enter or be immobilized at the water–chloroform interphase. The resulting high concentration of protein at the interphase causes protein to precipitate. Since the deproteinization action of chloroform occurs at the chloroform–water interphase, efficient deproteinization depends on the formation of a large interphase area. To achieve this, one has to form an emulsion of water and chloroform. Since chloroform does not mix with water this can only be done by vigorous shaking. An emulsifier, isoamyl alcohol, is added to chloroform to help form the emulsion and to increase the water–chloroform surface area.

The Marmur method is very efficient in the recovery of DNA, but it requires repeated time-consuming extractions when large amounts of protein are present. In addition, chloroform extractions require rather vigorous mixing that contributes to hydrodynamic shearing of large DNA molecules. Using this method, it is possible to obtain very pure DNA, but of limited size (20,000–50,000 bp). The method is useful for the preparation of DNA from viruses with small genomes or when DNA of low molecular weight is sufficient for experiments (e.g. a polymerase chain reaction or PCR).

A substantial improvement in the method can be accomplished by limiting the number of extractions. This saves time and limits DNA shearing. This can be done by enzymatically removing most of the protein before extraction. Another modification frequently used is combining phenol and chloroform extraction into one step.

The efficient use of the Kirby and Marmur methods of deproteinization of DNA requires prior enzymatic digestion of protein. These methods can only be used without this preliminary step when small amounts of protein contaminate DNA solutions.

Deproteinization using enzymes

Proteins can be removed from DNA preparations using a protease that can digest all proteins, i.e. a general-purpose protease. Two such enzymes are in use, proteinase K and pronase. Both enzymes are very stable, general specificity proteases that are secreted by fungi. Commercial preparations of these enzymes are inexpensive and devoid of DNase contamination, making them safe to use in the purification of nucleic acids. These proteases are active in the presence of low concentrations of anionic detergent, high concentrations of salts, and EDTA and exhibit broad pH (6.0–10.0) and temperature (50–67°C) optima. They can digest intact (globular) and denatured (polypeptide chain) proteins and do not require any co-factors for their activities. Proteinase K and pronase are usually used in DNA purification procedures at final concentrations of 0.1–0.8 mg ml^{-1}. The difference between these two enzymes lies in their activities towards self; pronase is a self-digesting enzyme, whereas proteinase K is not. The fact that proteinase K is not a self-digesting enzyme makes it a more convenient enzyme to use than pronase, because it is unnecessary to continually add it during the prolonged course of the reaction.

The major drawback in using these enzymes is that enzymatic treatment can only remove 80–90 percent of the proteins present. This is because protein digestion is an enzymatic reaction that is dependent on substrate and enzyme concentrations. In practice, the deproteinization rate depends only on the protein (substrate) concentration, because it is not practical to add a large amount of enzyme to accelerate the reaction at low substrate concentration. Therefore, as the reaction proceeds the concentration of substrate decreases progressively, thereby slowing the reaction rate and, indeed, enzymatic reactions will go to completion only given infinite time. At high substrate concentrations and sufficient concentration of enzyme, the reaction proceeds at a maximal rate until 80–90 percent of the substrate has been removed. The reaction rate then becomes too slow to be practical for the removal of remaining protein in a reasonable time.

The characteristics of enzymatic removal of proteins make enzymatic deproteinization an ideal and indispensable first step in nucleic acid purification. This treatment is used when a large amount of protein is present, i.e. right after cell lysis. The remaining proteins can be removed with a single extraction using organic solvent.

Removal of RNA

The removal of RNA from DNA preparations is usually carried out using an enzymatic procedure. Consequently this procedure does not remove all

RNA and, therefore, yields DNA preparations with a very small amount of RNA contamination. Two ribonucleases that can be easily and cheaply prepared free of DNase contamination are used, namely ribonuclease A and ribonuclease T1.

Ribonuclease A (RNase A) is an endoribonuclease that cleaves RNA after C and U residues. The reaction generates 2′:3′-cyclic phosphate which is hydrolyzed to 3′ nucleoside phosphate producing oligonucleotides ending with 3′-phosphorylated pyrimidine nucleotide.

Ribonuclease T1 (RNase T1) is an endoribonuclease that is very similar to RNase A in its reaction conditions and stability. The enzyme cleaves double-stranded and single-stranded RNA after G residues, generating oligonucleotides ending in a 3′-phosphorylated guanosine nucleotide.

Because of the RNA cleaving specificity of these enzymes, it is recommended that they be used **together** for complete RNA removal from DNA samples. The use of only one of these enzymes can result in contamination of DNA preparations with a large amount of oligonucleotides that will make the spectrophotometric measurement of DNA concentration practically impossible.

Concentrating the DNA

Precipitating with alcohol is usually performed for concentration of DNA from the aqueous phase of the deproteinization step. Two alcohols are used for DNA precipitation: ethanol and isopropanol.

Alcohol precipitation is based on the phenomenon of decreasing the solubility of nucleic acids in water. Polar water molecules surround the DNA molecules in aqueous solutions. The positively charged dipoles of water interact strongly with the negative charges on the phosphodiester groups of DNA. This interaction promotes the solubility of DNA in water. Ethanol is completely miscible with water, yet it is far less polar than water. Ethanol molecules cannot interact with the polar groups of nucleic acids as strongly as water, making ethanol a very poor solvent for nucleic acids.

Replacement of 95 percent of the water molecules in a DNA solution will cause the DNA to precipitate. Making a DNA solution of 95 percent ethanol is not practical because it requires the addition of a large volume of 100 percent ethanol. To precipitate DNA at a lower ethanol concentration, the activity of water molecules must be decreased. This can be accomplished by the addition of salts to DNA solutions. Moreover, the presence of salts will change the degree of charge neutralization of the DNA phosphates, eliciting extensive changes in the hydrodynamic properties of the DNA molecules (Eickbush and Moudrianakis, 1978). These changes, simultaneous with water elimination, will cause the separation of the DNA phase, i.e. precipitation, at the moment of complete neutralization of DNA molecules.

DNA precipitation is customarily carried out with 70 percent ethanol (final concentration) in the presence of the appropriate concentration of sodium or ammonium salts. The use of each of these salts has its advantages and disadvantages. The major advantage of using sodium chloride, in addition to convenience and low cost, is that SDS remains soluble in ethanol in the presence of 0.2 M NaCl. The use of sodium chloride is therefore recommended if a high concentration of SDS has been used for lysing the cells. The disadvantage of sodium chloride is its limited solubility in 70 percent ethanol making it difficult to completely remove from the DNA samples. This is particularly true when the precipitated DNA is collected by centrifugation. A high sodium chloride concentration in DNA preparations can interfere with the activity of many enzymes. When sodium chloride is used, the DNA should be spooled rather than centrifuged in order to collect precipitated DNA, making sodium chloride particularly useful in large-scale, high molecular weight DNA preparations.

Sodium acetate is more soluble in ethanol than sodium chloride and, therefore, is less likely to precipitate with the DNA sample. Its higher solubility in 70 percent ethanol makes it easier to remove from a DNA preparation by repeated 70 percent ethanol washes. Sodium acetate is the most frequently used salt in DNA precipitation.

Ammonium acetate is highly soluble in ethanol and easy to remove from precipitated DNA due to the volatility of both ammonium and acetate ions. The use of ammonium acetate instead of sodium acetate is also recommended for removing nucleotide triphosphates or small single- or double-stranded oligonucleotides (less then 30 bp), since these molecules are less likely to precipitate at high ethanol concentrations. In addition, precipitation of DNA with ammonium acetate has proven to be more efficient for the removal of heavy metals, detergents, and some unknown impurities that are potent inhibitors of restriction endonucleases and other enzymes used for DNA manipulation (Crouse and Amorese, 1987; Perbal, 1988).

Usually ethanol precipitation is carried out at temperatures of −20°C or lower. It is reasoned that low temperature and the presence of salts further lower the activity of water molecules, thereby facilitating more efficient DNA precipitation. However, a careful analysis of the efficiency of DNA precipitation at various temperatures and DNA concentrations demonstrated that this step could be performed at room temperature without serious loss of DNA, even when the concentration of DNA in a sample is very low (Zeugin and Hartley, 1985; Crouse and Amorese, 1987).

The best recoveries of DNA (DNA concentrations in the range 5–5,000 ng ml^{-1}) occur at room or 4°C temperatures and the worst when the precipitation is carried out at −70°C. The recovery of DNA at very low concentration (5 ng ml^{-1}) is not substantially different at the various temperatures and is largely dependent on time. Thus, in the procedures described in this manual DNA precipitation is always performed at room temperature.

Determination of the Purity and Quantity of DNA

The last step of any DNA isolation procedure is evaluation of the results. For DNA this evaluation involves determination of DNA concentration and evaluation of the purity of the DNA.

Ultraviolet (UV) spectrophotometry is used for the determination of DNA concentration. The DNA has maximum and minimum absorbances at 260 and 234 nm, respectively. However, these are strongly affected by the degree of base ionization and, hence, pH of the measuring medium (Beaven et al., 1955; Wilfinger et al., 1997).

The relationship between DNA absorbance at 260 nm (A_{260}) and DNA concentration (N) is described by the following equation:

$$N = A_{260}/\varepsilon_{260} \qquad (1.1)$$

where ε_{260} is the DNA extinction coefficient. This coefficient for double-stranded DNA is $0.02\,\mu g^{-1}\,cm^{-1}$ when measured at neutral or slightly basic pH. Thus, an absorbance of 1.0 at 260 nm gives a DNA concentration of $50\,\mu g\,ml^{-1}$ ($1/0.02 = 50\,mg\,ml^{-1}$). The value of the absorption coefficient (ε_{260}) for double-stranded DNA varies slightly depending on the percent of GC. As a result the concentration of DNA solutions having an absorbance of 1.0 is not always $50\,\mu g\,ml^{-1}$. This slight variation is usually disregarded. The absorption coefficient of single-stranded DNA is $0.027\,\mu g^{-1}\,cm^{-1}$ giving an ssDNA concentration of $37\,\mu g\,ml^{-1}$ for an absorbance of 1.0 ($1/0.027 = 37\,\mu g\,ml^{-1}$).

The linear relationship between absorbance at 260 nm and DNA concentration holds in a range between 0.1 and 2.0 absorbance units. Reliable measurements of DNA concentration can be made for solutions of 0.5–$100\,\mu g\,ml^{-1}$ using a standard UV spectrophotometer. Before measurement, samples with an absorbance equal to or greater than 2.0 should be diluted. The measurement of DNA concentration at a lower range (A_{260} lower than 0.2) can be strongly affected by light scattering on dust particles present in the preparation. Measuring the absorbance at 320 nm (Schleif and Wensink, 1981) will assess the degree of such contamination. At this wavelength, DNA does not absorb and any absorbance at 320 nm is due to light scattering. To measure DNA concentration properly the absorbance of a DNA sample at 320 nm should be less than 5 percent of the absorbance at 260 nm.

Absorbance measurements at wavelengths other than 260 nm are used for determination of the degree of protein contamination of the DNA sample. Proteins absorb maximally at 280 nm due to the presence of tyrosine, phenylalanine, and tryptophan and absorption at this wavelength is used for the detection of protein in DNA samples. This is done by determination of the $A_{260}:A_{280}$ ratio. This ratio for pure double-stranded DNA is not 1.8–1.9, as previously thought, but is 2.0 (Glasel, 1995; Huberman, 1995; Manchester, 1995; Held, 1997). The ratio between 1.8 and 1.9 corresponds to 60 and 40 percent protein contamination, respectively (Glasel, 1995). If the absorbance ratio of 260 nm : 280 nm is lower than 2.0, the DNA concentration can be calculated using following formula (Surzycki, 2000):

$$N(\mu g\ ml^{-1}) = 70\ A_{260} - 40\ A_{280} \tag{1.2}$$

where A_{260} and A_{280} are the absorbances of a DNA sample at 260 and 280 nm, respectively.

A better indicator of protein contamination in DNA samples is the ratio of $A_{260}:A_{234}$. DNA has an absorbance minimum at 234 nm and protein absorbance is high due to the absorption maximum for peptide bonds at 205 nm (Scopes, 1974; Stoscheck, 1990). Since the ratio of the DNA extinction coefficient at 234 nm (ε_{234}) to the protein extinction coefficient at the same wavelength is 1.5–1.8, the $A_{260}:A_{234}$ ratio is a very sensitive indicator of protein contamination. For pure nucleic acids, this ratio is between 1.8 and 2.0. The DNA concentration can be calculated from the absorbances at 260 and 234 nm using the following equation (Surzycki, 2000):

$$N(\mu g\ ml^{-1}) = 52.6\ A_{260} - 5.24\ A_{234} \tag{1.3}$$

where A_{260} and A_{234} are the absorbances of a DNA sample at 260 and 234 nm, respectively.

FIRST LABORATORY PERIOD

In this laboratory period you will collect your cheek cells and begin DNA purification.

Safety precautions

Each student should work only with his or her own cells. Any student who does not wish to isolate DNA from his or her own cells should be provided with human genomic DNA certified to be free of human immunodeficiency virus DNA. This DNA is commercially available from a number of companies (e.g. Promega Co. and Sigma Co.).

Special safety procedures are necessary when working with phenol or CIA solutions. Because of the relatively low vapor pressure of phenol, occupational systemic poisoning usually results from skin contact with phenol rather than from inhaling the vapors. Phenol is rapidly absorbed by and highly corrosive to the skin. It initially produces a white softened area, followed by severe burns. Because of the local anesthetic properties of the phenol, skin burns may not be felt until there has been serious damage. Gloves should be worn at all the times when working with this chemical. Because some brands of gloves are soluble or permeable to phenol, they should be tested before use. If phenol is spilled on the skin, flush off immediately with a large amount of water and treat with a 70 percent aqueous solution of PEG (polyethylene glycol) 4000 (http://users.ox.ac.uk/~phar0036/biomedsafety/labsafety/chemicalsafety/phenol.html). **Do not use ethanol.** Used phenol should be collected in a tightly closed, glass receptacle and stored in a chemical hood to await proper disposal.

The CIA reagent should also be handled with care. Mixing chloroform with other solvents can involve a serious hazard. Adding chloroform to a solution containing strong base or chlorinated hydrocarbons could result in an explosion. Prepare CIA in a fume hood because isoamyl alcohol vapors are poisonous. Store the CIA mixture in a hood in a tightly closed, dark glass bottle. Used CIA can be collected in the same bottle as phenol and discarded together.

Technical tips

Two of the most common obstacles in obtaining a high yield of high molecular weight DNA are hydrodynamic shearing and DNA degradation by non-specific DNases.

To avoid hydrodynamic shearing, DNA in solution should always be pipetted slowly with wide-bore pipettes (approximately 3–4 mm orifice diameter). A wide-bore pipette can be prepared by cutting off the tip of a 10 ml plastic disposable pipette. Alternatively, a pipette aid can be inserted on the tip end of a sterilized 10 ml glass pipette using the other end as the "intake" end. The end of the pipette should always be immersed in the liquid when pipetting DNA. The DNA solution should never be allowed to run down the side of a tube nor should it be vigorously shaken or vortexed.

In order to avoid DNase degradation, all solutions should contain DNase inhibitors. Two kinds of DNase inhibitors are in use, EDTA and detergents. EDTA is a Mg^{2+} ion chelator and a powerful inhibitor of DNases since most cellular DNases require Mg^{2+} ion as a co-factor for their activity. In addition, the presence of EDTA in extraction buffers inhibits Mg^{2+} ion-induced aggregation of nucleic acids. A concentration of 50–100 mM EDTA is usually sufficient for inhibiting DNases present in human cells. In fact, a high concentration of EDTA (above 100 mM) in buffers is not recommended because it leads to a substantial decrease in yields. Detergents commonly used in DNA purification are SDS or lithium deodecyl sulfate and Sarcosyl.

DNA samples should be stored under conditions that limit their degradation. Even at ideal storage conditions, one should expect approximately one phosphodiester bond break per 200 kb per year. For long-term storage, the pH of the buffer should be above 8.5 in order to minimize deamidation and contain at least 0.15 M NaCl and 10 mM EDTA.

During DNA preparation and storage the following conditions will contribute to fast degradation of DNA.

DNase contamination

The most frequent source of this contamination is human skin. In spite of the low stability of most DNases, even short exposure to a very low concentration of these enzymes will result in substantial sample degradation. In order to avoid this contamination, it is necessary to avoid direct or indirect contact between samples and fingers by wearing gloves and using sterilized solutions and tubes. Since DNA is easily absorbed onto glass surfaces only sterilized plastic tubes should be used for storage.

Presence of heavy metals

Heavy metals promote the breakage of phosphodiester bonds. Long-term DNA storage buffer should contain 10 mM or more of EDTA, which is a heavy metal chelator. If EDTA is present, DNA can be stored as a precipitate in 70 percent ethanol. This storage condition is preferred if the sample is stored at 5°C because it prevents bacterial contamination. A 1–2 mM EDTA concentration is sufficient for short-term storage of DNA and more convenient for everyday work.

Presence of ethidium bromide

The presence of ethidium bromide causes photo-oxidation of DNA with visible light in the presence of molecular oxygen. Since it is difficult to remove all ethidium bromide from DNA samples treated with this reagent, such samples should always be stored in the dark. Moreover, due to the ubiquitous presence of ethidium bromide in molecular biology laboratories, DNA samples can be easily contaminated with it. For this reason all DNA samples should be stored in the dark.

Temperature

The best temperature for short-term storage of high molecular weight DNA is between 4 and 6°C. At this temperature the DNA sample can be removed and returned to storage without cycles of freezing and thawing, which cause DNA breakage. For very long-term storage (five years or more) DNA should be stored at a temperature of −70°C or below, providing the sample is not subjected to any freeze–thaw cycles. Long- or short-term storage of high molecular weight DNA at −20°C is not recommended. This temperature can cause extensive single- and double-strand breakage of DNA because, at this temperature, molecular bound water is not frozen.

Protocol

Collecting human cheek cells

1. Pour 10 ml of PBS into a 15 ml conical centrifuge tube. Transfer the solution into a paper cup. Pour all the solution into your mouth and swish vigorously for 30–40 seconds. Expel the PBS wash back into the paper cup.
2. Transfer the solution from the paper cup into a 25 ml Corex centrifuge tube and place it on ice.
3. Repeat step 1 one more time with fresh PBS. Expel the mouthwash back into the paper cup and transfer the solution into the same 25 ml Corex tube.
4. Collect the cells by centrifugation at 5,000 r.p.m. for 10 minutes at 4°C.
5. Pour as much supernatant as possible back into the paper cup. Be careful not to disturb the cell pellet. Discard the supernatant from the paper cup into the sink. Invert the Corex centrifuge tube with cells on a paper towel to remove the remaining PBS.

DNA purification

1. Add 1 ml of lysis buffer prewarmed to 65°C to the cells and gently resuspend them by pipetting up and down.

2. Add 70 μl of proteinase K (20 mg ml^{-1}) and mix by inverting the tube several times.

3. Add 80 μl of 20 percent Sarcosyl and mix well by gently inverting the tube. Clearing of the "milky" cell solution and increased viscosity indicates lysis of the cells.

4. Incubate the mixture for 60 minutes in a 65°C water bath.

5. Add 800 μl of dilution buffer and mix carefully. Hold the tube between your thumb and index finger and very quickly invert several times. Do not allow the lysate to run slowly down the side of the tube. This step lowers the sodium chloride concentration allowing single- and double-stranded activities of both RNases.

6. Transfer the tube to a 37°C water bath and add 25 μl of RNase A and 2 ml of RNase T1. Cap the tube and mix by quickly inverting several times. Incubate for 30 minutes at 37°C.

7. Add 1 ml of phenol–8HQ solution and 1 ml of CIA solution to the tube. Close the tube with a Teflon-lined cap and mix using the procedure described in step 5. The solution should turn "milky" when properly mixed. This should not take more than two to three inversions.

8. Remove the cap and place the tube into a centrifuge. Centrifuge at 10,000 r.p.m. for 5 minutes at 4°C to separate the water and phenol phases. DNA will be in the top aqueous layer.

9. Collect the aqueous phase using a P1000 Pipetman set to 1,000 μl and equipped with a wide-bore blue tip. Prepare a wide-bore blue tip by cutting off 5–6 mm from the end of the tip with a razor blade. The total volume of the aqueous phase should be approximately 2 ml. Avoid collecting the white **powdery looking** precipitate at the interphase. However, do collect as much of the viscose, bluish-white layer from the interphase as possible. This layer contains concentrated nucleic acids, not proteins. Record the volume of aqueous phase and transfer it into a fresh 25 ml Corex tube. To do this place the cut-off end of the blue tip at the bottom of the tube and slowly deliver the solution.

10. Add 0.5 volume (1.0 ml) of 7.5 M ammonium acetate to the DNA solution and mix by inverting the tube.

11. Add approximately 6 ml of 95 percent ethanol to the tube containing DNA solution. The volume of added ethanol should be two times the total volume, i.e. DNA plus ammonium acetate (~3.0 ml). Carefully overlay the ethanol onto the viscose DNA solution. Since ethanol is less dense than the DNA solution, it will be the upper layer.

12. Precipitate DNA by gently inverting the tube several times. The DNA should appear as a cotton-like precipitate.

13. Insert the end of a glass hook into the precipitated DNA and swirl the hook in a circular motion to spool out the DNA. The DNA precipitate will adhere to the hook. **Note:** if at this step DNA does not form a clump and

instead it forms several smaller fragments, do not try to collect them on a glass hook. Go to step 17 instead.

14. Transfer the hook with DNA into a 20 ml tube filled with 5 ml of cold 70 percent ethanol. Wash the DNA by gently swirling the glass hook. Pour out the 70 percent ethanol and repeat the wash two more times.
15. Transfer the hook to a microfuge tube and add 300 μl of TE buffer. Rehydrate the DNA slowly by moving the glass hook back and forth. To speed up the rehydratation of DNA, incubate the solution in a 65°C water bath for 10–15 minutes moving the tube gently every 2–3 minutes.
16. Store the tube in a 4°C refrigerator until the next laboratory period
17. Collect precipitated DNA by centrifugation for 5 minutes at 10,000 r.p.m. Discard the supernatant by gently pouring off the ethanol. DNA will appear at the bottom of the tubes as a white precipitate.
18. Add 5 ml of cold 70 percent ethanol to the tube and wash the pellet by carefully rolling the tube at a 45° angle in the palm of your hand. Take care not to dislodge the DNA pellet from the bottom of the tube during this procedure. Never vortex the tube. Discard the 70 percent ethanol and drain the tube well by inverting it over a paper towel for a few minutes. Repeat the wash one more time.
19. Add 100 μl of TE buffer to the tube and rehydrate the DNA pellet by gently pipetting up and down using a P200 Pipetman equipped with a yellow tip with a cut-off end. Transfer the DNA solution into a 1.5 ml microfuge tube.
20. Add another 100 μl of TE buffer to the centrifuge tube and wash the tube by gently pipetting up and down as described in step 19. Add the solution to the microfuge tube with DNA. Repeat this washing one more time. The total volume of the DNA should be 300 μl. Store the tube at 4°C as described in step 16.

SECOND LABORATORY PERIOD

In this laboratory period we will continue DNA purification. First, DNA will be concentrated by precipitation with ethanol in the presence of ammonium acetate. We will use the ammonium acetate procedure because this salt has proven to be the most efficient for the removal of the heavy metals, detergents, and impurities that are potent inhibitors of restriction endonucleases and other enzymes used for DNA manipulation. Second, we will determine DNA concentration and purity using a UV spectrophotometer.

Safety precautions

The same safety precautions as for the first laboratory period apply.

Technical tips

The same technical tips as for the first laboratory period apply.

Protocol

Precipitating DNA

1. Retrieve the tube from the refrigerator and add 150 µl of 7.5 M ammonium acetate. Mix by inverting the tube several times.
2. Add 950 µl of 95 percent ethanol and mix by inverting the tube two to four times.
3. Place the tube in a centrifuge, orienting the attached end of the lid away from the center of rotation (see the icon in the margin). Centrifuge the tube at maximum speed for 5 minutes at room temperature.
4. Remove the tubes from the centrifuge. Pour off ethanol into an Erlenmeyer flask by holding the tube by the open lid and gently inverting the end. Touch the lip of the tube to the rim of the flask and drain the ethanol. You do not need to remove all of the ethanol from the tube. Return the tubes to the centrifuge in the same orientations as before. **Note:** when pouring off ethanol do not invert the tube more than once because this could disturb the pellet.

5. Wash the pellet with 700 µl of cold 70 percent ethanol. Holding the P1000 Pipetman vertically (see the icon in the margin) slowly deliver the ethanol to the side of the tube opposite the pellet. **Do not start the centrifuge:** in this step the centrifuge rotor is used as a "tube holder" that keeps the tube at an angle convenient for ethanol washing. Withdraw the tube from the centrifuge by holding the tube by the lid. Remove ethanol as before (step 4). Place the tube back into the centrifuge and wash with 70 percent ethanol one more time.

6. After the last ethanol wash, collect the ethanol remaining on the sides of the tube by centrifugation. Place the tubes back into the centrifuge with the side of the tube containing the pellet facing away from the center of rotation and centrifuge for 2–3 seconds. For this centrifugation, you do not need to close the lids of the tubes. Remove collected ethanol from the bottom of the tube using a P200 Pipetman equipped with capillary tip. **Note:** this procedure makes it possible to quickly wash the pellet without centrifugation and vortexing. Vortexing and centrifuging the pellet is time-consuming and leads to substantial loss of material and shearing of the DNA. Never dry the DNA pellet in a vacuum. This will make rehydration of the DNA very difficult if not impossible

7. Add 35 μl of TE buffer to each tube and resuspend the pelleted DNA. Use a yellow tip (P200 Pipetman) with a cut-off end for this procedure. Gently pipette the buffer up and down directing the stream of the buffer towards the pellet. If the pellet does not dissolve in several minutes, place the tube in a 60–65°C water bath and incubate for 10–20 minutes mixing occasionally.

Determination of DNA concentration and purity

1. Determine the concentration of DNA by measuring absorbance at 260 nm. Initially use a 1:20 dilution of the DNA. The absorbance reading should be in the range 0.1–1.5 OD_{260}. Special care must be taken to dilute the viscose solution of DNA when micropipettors are used. Most micropipettes will not measure the volume of a very viscose solution correctly. To prepare a 1:20 dilution of DNA, add 100 μl of PBS to a microfuge tube. Prepare a wide-bore, yellow tip by cutting off 5–6 mm from the end of the tip with a razor blade. Withdraw 5 μl of PBS from the tube and mark the level of the liquid with a marking pen. Discard PBS from the tip and draw DNA solution to the 5 μl mark. Transfer DNA to the tube containing PBS. Pipette up and down several times to remove the viscose DNA solution from the inside of the pipette tip. **Note:** DNA concentration should never be measured in water or TE buffer.

2. Determine the absorbance at 260 nm and calculate the DNA concentration using the equation

$$DNA (\mu g\ ml^{-1}) = OD_{260} \times 50 \times \text{Dilution factor} \tag{1.4}$$

3. Determine the purity of DNA by measuring the absorbances at 280 and 234 nm. Calculate 260 nm:280 nm and 260 nm:234 nm ratios. Calculate the amount of DNA using equations (1.2) and (1.3).

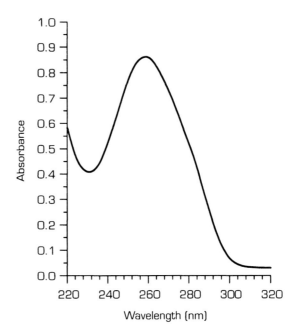

Figure 1.2 Absorption spectrum of DNA purified from human cheek cells. The DNA was diluted 20 times in PBS and scanned using a UV spectrophotometer. The 260 nm : 280 nm absorbance ratio was 1.7 and the 260 nm : 234 nm absorbance ratio was 1.8.

4. Label the tube with your name and group number and indicate the DNA concentration in **micrograms per microliter. Store the tube in a 4°C refrigerator.**

Expected results

A typical spectrum of the DNA purified from human cheek cells is shown in Fig. 1.2. The concentration of DNA isolated from two mouthwashes should be between 0.5 and 1.0 µg µl^{-1}. The total amount of DNA should be 17–35 µg. The DNA is high molecular weight and does not contain RNA.

A low 260 nm : 280 nm and/or 260 nm : 234 nm ratio indicates protein contamination and more often than not is caused by low activity of proteinase K or inadequate mixing of phenol and aqueous phases. The low activity of proteinase K is indicated by the presence of a large amount of "foamy" material at the interphase after the first phenol extraction, whereas inadequately mixed phenol and aqueous phases do not have a uniformly "milky" appearance.

The presence of low molecular weight DNA in the preparation frequently results from DNA mechanical shearing when pipettes with narrow openings are used or by allowing the DNA solution to run down the side of the tube. Inverting tubes too slowly during organic extraction will also result in substantial shearing of DNA molecules.

The presence of low molecular weight DNA in the preparation can also result from insufficient inhibition of DNase activity. This usually results from a too low concentration of EDTA in the lysis buffer.

A low yield of DNA can result from inadequate lysis of the cells or a too high concentration of EDTA in the lysis buffer. Insufficient cell lysis will be noticeable after the addition of Sarcosyl. Adequate lysis of the cells at this step results in a drastic increase in viscosity of the solution.

References

Beaven, G.H., Holiday, E.R., and Johnson, E.A. (1955) Optical properties of nucleic acids and their components. In *The Nucleic Acids*, Vol. 1, E. Chargaff and J.N. Davidson (eds), pp. 493–553. Academic Press, New York.

Crouse, J. and Amorese, D. (1987) Ethanol precipitation: ammonium acetate as an alternative to sodium acetate. *BRL Focus*, **9**(2), 3–5.

Eickbush, T.H. and Moudrianakis, E.N. (1978) The compaction of DNA helices into either continuous supercoils or folded-fiber rods and toroids. *Cell*, **13**, 295–306.

Glasel, J.A. (1995) Validity of nucleic acid purities monitored by 260 nm/280 nm absorbance ratios. *BioTechniques*, **18**, 62–3.

Held, P. (1997) The importance of 240 nm absorbance measurement. The A_{260}/A_{280} ratio just isn't enough anymore. *Biomed. Products*, **7**, 123.

Huberman, J.A. (1995) Importance of measuring nucleic acid absorbance at 240 nm as well as at 260 and 280 nm. *BioTechniques*, **18**, 636.

Kirby, K. (1957) A new method for the isolation of deoxyribonucleic acids; evidence on the nature of bonds between deoxyribonucleic acid and proteins. *Biochem. J.*, **66**, 495–504.

Manchester, K.L. (1995) Value of A_{260}/A_{280} ratios for measurement of purity of nucleic acids. *BioTechniques*, **19**, 208–10.

Marmur, J. (1961) A procedure for the isolation of deoxyribonucleic acid from microorganisms. *J. Mol. Biol.*, **3**, 208–18.

Perbal, B.V. (1988) *A Practical Guide to Molecular Cloning*, 2nd edn. John Wiley & Sons, New York, Chichester, Brisbane, Toronto, and Singapore.

Schleif, R.F. and Wensink, P.C. (1981) *Practical Methods in Molecular Biology*. Springer-Verlag, New York.

Scopes, R.K. (1974) Measurement of protein by spectrophotometry at 205 nm. *Anal. Biochem.*, **59**, 277–82.

Stoscheck, C.M. (1990) Quantitation of protein. *Methods Enzymol.*, **182**, 50–68.

Surzycki, S.J. (2000) *Basic Techniques in Molecular Biology*. Springer-Verlag, Berlin, Heidelberg, and New York.

Wilfinger, W.W., Mackey, K., and Chomczynski, P. (1997) Effect of pH and ionic strength of the spectrophotometric assessment of nucleic acid purity. *BioTechniques*, **22**, 474–80.

Zeugin, J.A. and Hartley, J.L. (1985) Ethanol precipitation of DNA. *BRL Focus*, **7**(4), 1–2.

CHAPTER 2

DNA Fingerprinting: Multi-locus Analysis

Introduction

The goal of this experiment is to carry out multi-locus DNA fingerprinting. For this procedure you will use your own DNA and a probe that recognizes a family of DNA mini-satellites with the core sequence GAGGGTGGNG GNTCT. The typing of your DNA will require performing five procedures. The essential steps of the typing procedure are as follows.
1. DNA is digested into fragments using restriction endonuclease enzymes. We will use the enzyme *Hae*III, which is widely used in forensic work. Experiment 1 describes this procedure.
2. DNA fragments resulting from digestion are separated based on size by agarose gel electrophoresis. Experiment 2 describes this technique.
3. DNA fragments are transferred from gel onto a nylon membrane by a process termed Southern blotting. The membrane will contain the DNA fragment located in exactly the same position as it was present in the gel. This procedure will be performed in experiment 3.
4. Immobilized DNA fragments are hybridized with labeled DNA probe complementary to core mini-satellite sequences. Labeling of the probe and hybridization will be carried out in experiment 4.
5. Hybridized DNA fragments are detected by chemiluminescence. This is described in the signal detection experiment.
The entire experiment will be done during five laboratory periods. Figure 2.1 presents the overall timetable for these experiments.

Background

The human genome contains approximately 3 billion bp. This genome, similar to all other higher eucaryotes, may be divided into classes based very broadly on their functional properties. Approximately 10 percent of the genome constitutes DNA sequences harboring genetically relevant

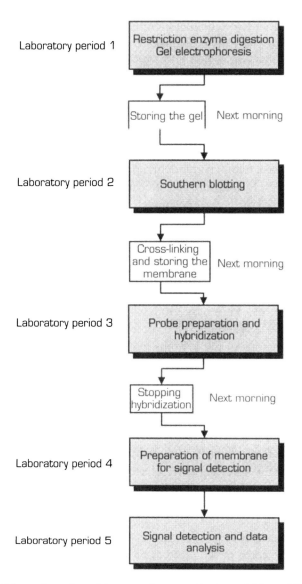

Figure 2.1 Schematic outline of the procedures used in the multi-locus DNA fingerprinting experiment.

information that is essential for each individual, i.e. the gene-coding region of the DNA. The remaining DNA constitutes non-coding regions of the genome. This part of the genome, largely due to its non-coding nature, is the major source of variability or polymorphism-responsible differences between individuals. The non-coding segment of the genome consists of two classes of DNA sequences: unique low copy number DNA and repetitive DNA. Repetitive DNA can in turn be further subdivided into interspersed repetitive DNA and tandemly repeated DNA.

Tandemly repetitive sequences, commonly known as "satellite DNAs" are classified into three major groups.

1. **Satellites.** These are very highly repetitive DNA sequences with repeat lengths of one to several thousand base pairs. These sequences are typically organized as large clusters (up to 100 million bp) in the heterochromatic regions of chromosomes, centromeres, and telomeres. Human centromeric and pericentromeric DNA consists predominantly of satellites II, III, IV, alphoid, and I that are specific or nearly specific for particular chromosomes. In contrast, telomeric tandem repeats are conserved and identical for all chromosomes. In addition, specific satellite DNA sequences are present on the long arm of the Y chromosome.

2. **Micro-satellites.** These are moderately repetitive DNA sequences composed of arrays of short repeats (2–10 bp). The human genome contains at least 30,000 micro-satellite loci located in euchromatin. The number of repeats is characteristically variable within a population for each micro-satellite, typically with mean array sizes in the order of ten to 100 repeats.

3. **Mini-satellites.** These are moderately repetitive, tandemly repeated arrays of 10–100 bp spanning 0.5 kb to several kilobases. They are found in euchromatic regions of the genome and are predominantly clustered towards chromosome ends. They are highly variable in array size.

In general, satellite DNAs can be variable among individuals and, thus, form excellent tools for genetic individualization, particularly with regard to the number of repeats at a given locus. Mini-satellite loci are the most highly polymorphic sequence elements yet discovered in the human genome and delineating the repeat lengths of these loci is the basis of most DNA typing systems used in forensic medicine. These loci are usually referred to as variable number tandem repeat (VNTR) loci. The VNTRs can be grouped into families of independent loci that are related to each other by small variation in their common core sequence. Some VNTR loci are hypervariable and contain between 100 and 1,000 repeats. The variability of these loci is not limited to differences in the number of the repeated unit, but also the sequence of the repeat can vary in different members of an array. Thus, any given hypervariable VNTR allele can be monomorphic for length, but may still be polymorphic in structure. At present approximately 300 human mini-satellite families have been typed and less than ten of them are hypervariable (Nakamura et al., 1987; Armour et al., 1990; Amarger et al., 1998; Vergnaud and Denoeud, 2000).

Hypervariable VNTR loci are used in genetic typing by means of two methodologies: a **multi-locus analysis** or DNA fingerprinting and a **single-locus analysis** or single-locus DNA typing. Jeffreys et al. (1985a, b) first introduced DNA fingerprinting for individual identification in 1985. Soon after this the first case, which involved a UK immigration dispute, was satisfactorily resolved by DNA fingerprinting. Shortly after the method was used in an unusual paternity dispute in a UK court. DNA fingerprinting

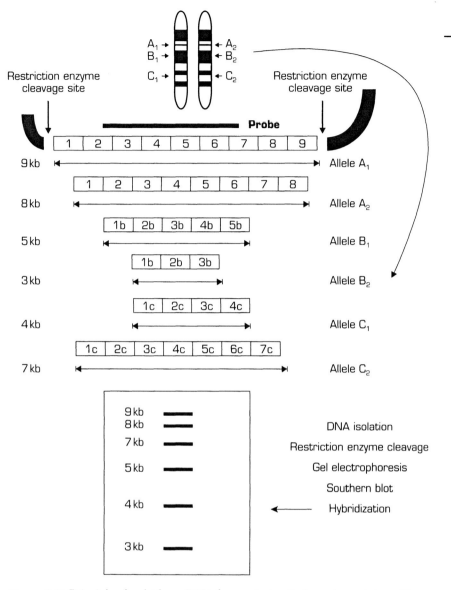

Figure 2.2 Principle of multi-locus DNA fingerprinting. Independent mini-satellite loci A, B, and C are members of one VNTR family related to each other by a small variation in the common core sequence of 1,000 bp. Each locus has a different number of repeats on each homologous chromosome, designated as A_1 and A_2, B_1 and B_2, and C_1 and C_2. In order to generate a DNA fingerprint, DNA is cut with restriction endonuclease that does not have a recognition site on any repeat. This generates a set of DNA fragments of different sizes, which is a consequence of the different number of repeats present at a particular locus. Thus, locus A_1 will be represented by 9 kb fragments, locus A_2 by 8 kb fragments, locus B_1 by 5 kb fragments, etc. These fragments are separated by agarose gel electrophoresis, transferred to a membrane, and hybridized to a probe complementary to the repeated element. The autogram shows a set of hybridization bands that represent hybridization to each member of the VNTR family. This is called a DNA fingerprint.

made its debut in a criminal case in 1986 in the Enderbery murder case (Jeffreys and Pena, 1993). After a period of initial disputes as to the validity of DNA fingerprinting for unequivocal identification of an individual, the method became established worldwide in forensic medicine and criminal investigations (Benecke, 1997). DNA typing has found many uses other than forensic applications (Kirby, 1990). These include animal and plant breeding, conservation biology, patent identification, genototoxicity studies, etc.

Both, DNA typing (single-locus analysis) and DNA fingerprinting (multi-locus analysis) use the same technique for revealing the polymorphic variation in the number of tandem repeats. DNA is digested with restriction enzyme that cuts externally to the tandem repeat, fragments are separated by gel electrophoresis, and Southern blots are hybridized to a probe either recognizing an internal core sequence (DNA fingerprinting) or a locus-specific sequence (single-locus analysis). The principle of multi-locus DNA fingerprinting is presented in Fig. 2.2.

The quantity of non-degraded DNA needed for these analyses is in the order of 1–10 µg. Because, with the exception of parentage studies, the quantity and integrity of DNA in a typical forensic specimen is limited, a PCR (polymerase chain reaction)-based technique was developed. This technique requires only a few nanograms of DNA and, indeed, the DNA of a single cell is usually sufficient for successful DNA typing. An efficient PCR is possible for fragments of approximately 1 kb in size. Most classical VNTR alleles are much longer than that limit and cannot be reliably amplified. However, PCRs can be used for amplifying tandem repeats of microsatellites, named short tandem repeats (STRs), for which the repeat arrays are in the range of 50–300 bp. At present forensic DNA analysis uses STR loci exclusively for individual identification. These loci are composed of tri-, tetra-, and pentameric core units and are evenly distributed throughout all chromosomes. Because the primer used is unique for a given STR locus, this method uses a single-locus approach for data analysis.

FIRST LABORATORY PERIOD

In this laboratory period two experiments will be performed. First, you will digest your DNA with restriction enzyme endonuclease. Second, you will carry out agarose gel electrophoresis of restricted DNA.

Experiment 1: restriction enzyme digestion

Introduction

In this experiment you will digest DNA purified in Chapter 1 using type II restriction enzyme endonuclease *Hae*III. You will digest your DNA and your partner's DNA as well as two control DNAs that will be given to you by your instructor. This experiment will take a single laboratory period.

The *Hae*III restriction enzyme endonuclease is the most often used restriction endonuclease in human DNA fingerprinting. The recognition sequence for this enzyme is GGCC assuring extensive digestion of human DNA. This recognition site is not present in the M13 15 bp tandem repeat element (GAGGTGGNGGNTCT) that we will use for the multi-locus fingerprint of your DNA.

Background

Restriction endonucleases catalyze sequence-dependent, double-stranded breaks in DNA yielding a homogeneous population of DNA fragments. These enzymes are used in a number of applications in molecular biology, including establishment of an endonuclease map of DNA, fragmentation of genomic DNA prior to Southern blotting, generation of fragments that can be subcloned in appropriate vectors, and generation of fragments for labeled probes.

The most frequently used restriction endonucleases belong to the type II endonucleases discovered by Smith and co-workers (Smith and Wilcox, 1970). These enzymes are small, monomeric proteins that require only Mg^{2+} for activity. Type II enzymes recognize a short nucleotide sequence with dyad symmetry (the 5′ to 3′ nucleotide sequence of one DNA strand is identical to that of the complementary strand sequences). Most sites consist of 4, 5, or 6 bp (Brooks, 1987), but a few have a recognition site of 8 bp or larger or sites smaller than 4 bp (Roberts and Macelis, 1991). In general, there are three possible cleavage positions within a recognition sequence: at the center of the axis of symmetry, yielding "flush" or "blunt" ends (e.g. CCC3′|5′GGG), to the left of the center giving cohesive termini with a protruding 5′-phosphate (e.g. C3′|5′CGG), or to the right of the center giving cohesive termini with protruding 3′-phosphates (e.g. CTGCA3′|5′G).

An estimate of the number of cleavage sites for a restriction endonuclease within a given piece of DNA, assuming an even distribution of bases, is described by the equation site number = $N/4^n$, where N is a number of base pairs in the DNA and n is the number of bases in the recognition site of the restriction endonuclease. This should be treated only as an approximation of the number of expected sites (Rodriguez and Tait, 1983).

Owing to the complicated nature of the restriction reaction and differences in substrates, it is difficult to define universal units of activity for these enzymes. For this reason, a convention was adopted for defining a unit of enzyme activity as the amount of enzyme required to digest 1 µg of bacteriophage lambda DNA completely in 1 hour.

Despite the diversity of the sources and of the specificity of over 1,000 type II restriction endonucleases identified to date, restriction enzyme reaction conditions are remarkably similar. Each restriction enzyme has its own optimal reaction conditions, which are usually given on the information sheet provided by the manufacturer. The major variables are the temperature of incubation and the composition of the buffer. The temperature requirements of type II restriction enzymes are very strict, whereas the differences between salt and pH requirements are often only slight. Presently the manufacturer of the enzyme supplies the appropriate buffer for each enzyme.

Technical tips

The general rules for working with restriction enzymes and preparing a digestion reaction are as follows.

1. Store restriction endonuclease at −20°C in a freezer that is not frost free at a concentration of 10 u µl^{-1} or higher.

2. The volume of the digestion reaction should be large enough that the restriction enzyme constitutes no more than 10 percent of the total volume. A 30 µl reaction volume should be routinely used.

3. Use a DNA amount no greater than 10 µg added in a volume that is not to exceed one-third of the reaction volume. Addition of a large volume of DNA dissolved in TE buffer will decrease the Mg^{2+} ion concentration in the reaction, thereby inhibiting restriction enzyme activity.

4. Use ten units or more of enzyme per microgram of DNA. Although this is far more enzyme than is theoretically required, this excess assures complete digestion in the case of impurities in the DNA, decreased enzyme activity from storage, pipetting errors during enzyme addition, etc. Some enzymes cleave their defined sites with different efficiency, largely due to differences in the flanking nucleotides and cleavage rates. Different sites recognized by a given enzyme can differ by a factor of ten. Using excess enzyme does compensate for these differences. Use 20-fold excess enzyme when digesting human genomic DNA.

Protocol

Always observe the following rules when preparing the reaction mixture. First, thaw all reagents at room temperature and place them on ice. Second, calculate the amounts of all reagents needed. Do not include water in these calculations. Next calculate the amount of water needed to obtain the desired reaction volume.

Start the assembly of the reaction mixture by addition of water. Remember the rule for reaction mixture assembly: **the amount of water is always calculated last, but water is always added first.** Add the remaining ingredients in the following order: buffer, co-factors, and substrate. Start the reaction by the addition of enzyme.

1. Label four sterile 1.5 ml microfuge tubes as follows: 1H, 2H, 3H, and 4H.
2. Calculate the volume of DNA solution to be drawn in order to have 3 µg of DNA per reaction. Use your DNA concentration as determined in the DNA isolation experiment. Your instructor will give you the concentrations of control DNAs. Record the results of these calculations in Table 2.1.
3. Calculate the amount of enzyme and buffer to be added to each reaction. The concentration of the buffer stock solution is ten times and the enzyme concentration in the stock solution is $10\,u\,\mu l^{-1}$. You will need 30 units of enzyme per reaction. Record these calculations in Table 2.1.
4. Calculate the amount of water needed to obtain a final reaction volume of 30 µl.
5. Add the calculated amount of water to each tube.
6. Add 3 µl of ten times buffer to each tube. Mix by pipetting up and down.
7. Add the appropriate DNA for each tube. Consult Table 2.1 for the amount and type of DNA to be added to each tube. For the addition of each DNA you will need to prepare a wide-bore yellow tip. Prepare each tip by cutting off 5–6 mm from the end of the tip with a razor blade. Set a P20

Table 2.1 Restriction enzyme digestion of DNA

Tube number	1H	2H	3H	4H
Buffer (ten times)	3.0 µl	3.0 µl	3.0 µl	3.0 µl
DNA control 1	–	–	–	–
DNA control 2	–	–	–	–
Your DNA	–	–	–	–
Partner's DNA	–	–	–	–
HaeIII enzyme ($10\,u\,\mu l^{-1}$)	3.0 µl	3.0 µl	3.0 µl	3.0 µl
Water				
Total	30.0 µl	30.0 µl	30.0 µl	30.0 µl

Pipetman to the required volume for each DNA type. Draw that amount of TE buffer into the tip and mark the level of the liquid with a marking pen. Discard TE buffer from the tip. Using the same tip draw the DNA solution to this mark and transfer it to the reaction mixture. Pipette up and down several times in order to remove the viscose DNA solution from the inside of the pipette tip. **Note:** it is important to follow the procedure described above in order to prevent DNA shearing during pipetting.

8. Start reactions with the addition of enzyme. Mix the enzyme with the reaction mixture by pipetting up and down several times.

9. Centrifuge the tube for 5–10 seconds in order to remove air bubbles and collect liquid at the bottom of the tube.

10. Transfer the tubes to a 37°C water bath and incubate for 1–2 hours.

11. Begin preparing the agarose gel as described in the protocol of step 2 in experiment 2.

References

Amarger, V., Gauguier, D., Yerle, M., Apiou, F., Pinton, P., Giraudeau, F. et al. (1998) Analysis of the human, pig, and rat genomes supports a universal telomeric origin of minisatellite sequences. *Genomics*, **52**, 62–71.

Armour, J.A.L., Povey, S., Jeremiah, S., and Jeffreys, A.J. (1990) Systematic cloning of human minisatellites from orders array charomic libraries. *Genomics*, **8**, 501–12.

Benecke, M. (1997) DNA typing in forensic medicine and in criminal investigations: a current survey. *Naturwissenschaften*, **84**, 181–8.

Brooks, J.E. (1987) Properties and uses of restriction endonucleases. *Methods Enzymol.*, **152**, 113–29.

Jeffreys, A.J. and Pena, S.D.J. (1993) Brief introduction to human fingerprinting. In *DNA Fingerprinting. State of the Science*, S.D.J. Pena, R. Chakraborty, J.T. Epplen, and A.J. Jeffrey (eds), pp. 1–20. Birkhauser Verlag, Basel, Boston, and Berlin.

Jeffreys, A.J., Wilson, V., and Thein, S.L. (1985a) Individual-specific "fingerprints" of human DNA. *Nature*, **316**, 76–9.

Jeffreys, A.J., Brookfield, J.F.Y., and Semeonoff, R. (1985b) Positive identification of an immigration test-case using human DNA fingerprinting. *Nature*, **317**, 818–19.

Kirby, L.T. (1990) *DNA Fingerprinting. An Introduction*. Stockton Press, New York.

Nakamura, Y., Leppert, M., O'Connell, P., Wolff, R., Holm, T., Culver, M. et al. (1987) Variable number of tandem repeat (VNTR) markers for human gene mapping. *Science*, **235**, 1616–22.

Roberts, R.J. and Macelis, D. (1991) Restriction enzymes and their isoschizomeres. *Nucleic Acids Res.*, **19** (Suppl.), 2077–109.

Rodriguez, R.L. and Tait, R.C. (1983) *Recombinant DNA Techniques: An Introduction*. Addison-Wesley Publishing Co., Reading, MA.

Smith, H.O. and Wilcox, K.W. (1970) Restriction enzymes from *Hemophilius influenzae*. Purification and general properties. *J. Mol. Biol.*, **51**, 379–91.

Vergnaud, D. and Denoeud, F. (2000) Minisatellites: mutability and genome architecture. *Genome Res.*, **10**, 899–907.

Experiment 2: agarose gel electrophoresis

Introduction

The procedure describes the use of a large agarose gel in the separation of human DNA fragments. DNA band separation in such gels is sufficient for detecting a single copy gene in a human genome. These gels are well suited for Southern blotting. In order to attain higher resolution, electrophoresis will be run at a low voltage gradient of $1\,\mathrm{V\,cm^{-1}}$. Thus, the electrophoresis time will be longer than usual, but the resolution of DNA fragments will be better, in particular for higher molecular weight DNA.

Background

Principle of electrophoresis

When a molecule is placed in an electric field it will migrate to the appropriate electrode with a velocity or free electrophoretic mobility (M_0), which is described by the equation:

$$M_0 = \frac{E}{d} \frac{q}{6\pi R \eta} \qquad (2.1)$$

where E is the potential difference between electrodes measured in volts, q is the net charge of the molecule, d is the distance between electrodes (cm), η is the viscosity of the solution, R is Stock's radius of the molecule, and E/d is the field strength.

Since under physiological conditions phosphate groups in the phosphosugar backbone of DNA (RNA) are ionized, these polyanions will migrate to the positive electrode (anode) when placed in an electric field. Due to the repetitive nature of the phosphosugar backbone, double-stranded DNA molecules have a net charge to mass ratio that is approximately the same. Consequently, DNA molecules have approximately the same free electrophoretic mobility (M_0) irrespective of their size. It is apparent from equation (1) that the effects of friction on the mobility of the molecules can be accentuated by changing the viscosity (η) of the electrophoretic medium. If the viscosity is very large, the mobility of the molecules subjected to electrophoresis will depend largely on their shape and size. Equation (2.1) simplifies to:

$$M_0 = \frac{E}{d} \frac{1}{R} \qquad (2.2)$$

Specific support matrixes are used for increasing the viscosity of an electrophoretic medium. These include agarose and polyacrylamide. Varying the pore size using various agarose concentrations or different cross-linking ratios of polyacrylamide alters the viscosity of these materials. The mobility of DNA molecules is profoundly influenced by the size and shape of the molecules, as well as by the size of the matrix pores. Using these gels, DNA molecules are fractionated by their size and conformation in a relatively fast and inexpensive way.

Principle of agarose gel electrophoresis of DNA

Agarose is a polysaccharide consisting of basic agarobiose repeat units of 1, 3-linked β-D-galactopyranose and 1, 4-linked 3, 6-anhydro-α-L-galactopyranose. Units form long chains of approximately 400 repeats, reaching a molecular weight of approximately 120,000 Da. Long polymer chains contain small amounts of charged residues consisting largely of pyruvate and sulfate that are responsible for agarose properties that are important in gel electrophoresis. To these belongs the phenomenon of electroendosmosis (Adamson, 1976; Hiemenz, 1977).

During electrophoresis only hydrated positive ions, which are normally associated with the fixed anionic groups of agarose (pyruvate or sulfate residues), can move towards the cathode. Water is therefore pulled with these positive ions towards the negative electrode and negative molecules, such as DNA migrating towards the positive electrode, are slowed down. Thus, for maximum separation of DNA molecules by agarose gel electrophoresis, agarose with the lowest possible electroendosmosis should be used.

The electrophoretic migration rate of DNA through agarose gel depends on the following parameters: the size of the DNA molecules, the concentration of agarose, the voltage applied, the conformation of the DNA, and the buffer used for electrophoresis.

On first approximation, DNA molecules travel through gel at a rate inversely proportional to the logarithm of their molecular weight or number of base pairs. Therefore, a plot of mobility against the log of the size should give a straight line for all DNA sizes. However, this is true for a narrow size range. A better linear relationship between mobility and DNA size is obtained in plots of DNA base pair number (DNA size) versus 1/mobility (Hiemenz, 1977; Sealey and Southern, 1982).

The useful linear range of mobility depends on the gel concentration used and voltage applied. A DNA fragment of a given size migrates at different rates in gels containing different concentrations of agarose. A model for gel structure predicts that the log of the mobility of different DNA molecules (M) as a function of gel concentration (C) should result in a straight line with different slopes called retardation coefficients (K_r) and intercepts the so-called free mobility (M_0), i.e. the mobility of DNA molecules at zero concentration of agarose. This can be expressed mathematically by the following equation:

$$\log M = \log M_0 - CK_r \qquad (2.3)$$

It is possible to resolve a wide range of DNA fragment sizes using gels of different agarose concentrations provided that the voltage gradient applied to the gel is chosen correctly. Normally the migration rate of DNA fragments is directly proportional to the voltage applied. However, with increased voltage large DNA molecules migrate at a rate proportionally faster than small molecules. Consequently, the field strength applied to most gels should be between 0.5 and $10\,V\,cm^{-1}$. In general, higher resolution is achieved at a low voltage gradient, particularly if higher molecular weight DNA is used.

The amount of DNA in a sample will also affect its apparent mobility. Overloaded bands will appear to move faster than bands with the correct amount of DNA. For that reason, the amount of DNA loaded should be similar when comparing the mobility of DNA fragments. The useful separation ranges of various gel concentrations do overlap and different electrophoretic conditions can shift their useful range (see Surzycki (2000) for details).

Electrophoresis buffers

Several different buffers are used for agarose gel electrophoresis. These are TAE (Tris-acetate EDTA) buffer, TBE (Tris-borate EDTA) buffer, and TPA (Tris-phosphate EDTA) buffer.

TAE buffer is the most frequently used buffer for DNA electrophoresis. This buffer has a rather low buffering capacity, but permits the application of high-voltage gradients resulting in shorter running times. The ratio of voltage applied to current (mA) is approximately 1.0 for a wide variety of gel sizes and buffer volumes when this buffer is used (Perbal, 1988). The tracking dye, bromophenol blue, will travel in this buffer at a rate of approximately $1\,cm\,h^{-1}$ at $1-10\,V\,cm^{-1}$ field strength. Thus, this marker dye co-migrates with the smallest DNA molecules at each agarose concentration.

TBE buffer has a very high buffering capacity. It can be used when DNA of less than 12,000 bp is electrophoresed, but gives superior results, as compared to TAE buffer, in electrophoresis of DNA fragments of less than 1,000 bp. The DNA mobility in this buffer is approximately two times slower than in TAE buffer. This is due to the lower porosity of agarose gel when agarose polymerizes in the presence of borate. The ionic strength of TBE buffer is high, resulting in a 4:1 ratio of voltage to current (mA) during electrophoresis for a wide variety of gel sizes and buffer volumes (Perbal, 1988). In general, DNA bands are sharper when TBE buffer is used, but the time of electrophoresis is considerably longer.

TPA buffer has high buffering capacity, comparable to that of TBE buffer. However, the DNA mobility in this buffer is similar to that in TAE buffer due to a similar pore size formed during agarose polymerization. The buffer has a high ionic strength resulting in a voltage to current (mA) ratio similar to that obtained in TBE buffer.

Gel size

Agarose gel electrophoresis is commonly carried out using submerged horizontal slab gels (submarine gels). The best separation between DNA bands in such a system is achieved in gels that are approximately 20 cm long, 15 cm wide, and approximately 4 mm thick. To obtain maximum resolution of many bands electrophoresis should be continued until the tracking dye (for example bromophenol blue) has moved 70–80 percent of the length of the gel.

The size of the sample well can also affect the resolution of DNA bands. The optimal length of the sample well for a large gel is between 0.5 and 1 cm and the optimal width of the well is 1–2.0 mm. The sample well bottom should be 0.5–1.0 mm above the gel bottom. Most of the commercially available submarine electrophoresis gel boxes fulfill the above requirements.

Sample concentration

The amount of DNA loaded into one well can vary considerably without affecting the mobility of the DNA. For standard large gels, the DNA load can vary from 1 to 10 ng of DNA per band. The total amount of DNA loaded per well should not exceed 10 µg. In general, decreasing amounts of DNA should be used with increasing voltage.

Sample loading solutions

DNA samples are prepared for electrophoresis by the addition of loading dye solution. The composition of loading dye solution plays an essential role in obtaining sharp DNA bands. This solution serves three vital functions: it is used to terminate enzymatic reactions before electrophoresis (stop solution), it provides density for loading the sample into the well, and it provides a way of monitoring the progress of electrophoresis.

Most loading dye solutions contain EDTA in order to stop enzymatic reactions. However, this is frequently not sufficient for fully dissociating DNA–protein complexes, the presence of which will affect not only the mobility of DNA fragments, but can also cause an excessive smearing and widening of the bands. In order to remove these complexes, loading dye solutions should contain a protein-denaturing agent. Urea, at a concentration of 5 M, is the best protein-denaturing agent because it does not interact with agarose or affect DNA mobility.

Glycerol or sucrose, at concentrations of 5–10 percent, is used to provide the sample density for loading. However, using these low molecular weight compounds results in U-shaped DNA bands due to sample streaming up the side of the well before beginning electrophoresis (Sealey and Southern, 1982). This effect is particularly pronounced when electrophoresis is carried out at low field strength. To increase the sharpness of the bands and prevent their U-shape appearance, Ficoll 400 should be added at a concentration of 15–20 percent.

Gel staining

In order to visualize DNA, agarose gels are usually stained with ethidium bromide. This is the most rapid, sensitive, and reproducible method currently available for staining single- and double-stranded DNA (Sharp et al., 1973). Ethidium bromide binds to double-stranded nucleic acid by intercalation between stacked base pairs. The mobility of linear DNA in gel electrophoresis is reduced by approximately 15 percent. Ultraviolet (UV) irradiation of ethidium bromide at 302 and 366 nm is absorbed and re-emitted as fluorescence at 590 nm. Similarly, energy absorbed by DNA irradiated at 260 nm is transmitted to intercalated dye and re-emitted as fluorescence at 590 nm. The intensity of fluorescence of dye bound to DNA is much greater than that of free dye suspended in agarose. This results in very low background and a high intensity of fluorescence from DNA bands. The best staining results are obtained by incorporating ethidium bromide into the gel at a concentration of $0.5\,\mu g\,ml^{-1}$. This permits direct observation of the progress of electrophoresis and limits the amount of ethidium bromide-contaminated liquid waste. The most sensitive photographs of ethidium bromide-stained DNA are obtained when DNA is illuminated with UV light at 254–260 nm rather than by ethidium bromide direct illumination at 300 nm.

Photographing gels

Photographs of the gels provide not only a permanent record of the experiment, but also permit analysis of the data and visualization of DNA bands not visible to the unaided eye. Polaroid cameras, equipped with appropriate filters, are usually used for this purpose. The most commonly used fast Polaroid film type 667 (ASA 3000) can record a DNA band containing 2–4 ng when loaded into a well of 1 cm.

A more sensitive method of recording gel results is the use of a computer imaging system equipped with a charge-couple device digital camera. The sensitivity of the computer imaging system is approximately ten times greater than the sensitivity of photography. This permits visualization and recording of as little as 0.1 ng of DNA per band.

Safety precautions

Ethidium bromide is a mutagen and suspected carcinogen. Contact with skin should be avoided. Wear gloves when handling ethidium bromide solution and gels containing ethidium bromide.

For safety purposes, the electrophoresis apparatus should always be placed on the laboratory bench with the positive electrode (red) facing away from the investigator, that is away from the edge of the bench. To avoid electric shock always disconnect the red (positive) lead first.

UV light can damage the retina of the eye and cause severe sunburn. Always use safety glasses and a protective face shield to view the gel. Work in gloves and wear a long-sleeved shirt or laboratory coat when operating UV illuminators.

Protocol

1. Stop restriction reactions by the addition of 5 µl of stop solution. Mix well by pipetting up and down several times. Centrifuge for 5–10 seconds to collect liquid at the bottom of the tube. The reaction is now ready to be loaded onto an agarose gel as described in experiment 2. **Note:** restriction digestion can also be stopped by the addition of 1 µl of 0.5 M EDTA. The reaction can be stored at −20°C for an indefinite time and used for gel electrophoresis when needed.

2. Seal the opened ends of the gel-casting tray with tape. Regular labeling tape or electrical insulation tape can be used. Check that the teeth of the comb are approximately 0.5 mm above the gel bottom. To adjust this height, it is most convenient to place a plastic charge card (e.g. MasterCard) under the comb and adjust the comb height to a position where the card is easily removed from under the comb.

3. Prepare 1,500 ml of one times TAE buffer by adding 30 ml of a 50 times TAE buffer stock solution to a final volume of 1,500 ml of deionized water.

4. Place 150 ml of the buffer into a 500 ml flask and add the appropriate amount of agarose. Weigh 1.5 g of agarose for a 1 percent agarose gel. Melt the agarose by heating the solution in a microwave oven at full power for approximately 3 minutes. Carefully swirl the agarose solution to ensure that the agarose is dissolved, that is no agarose particles are visible. If evaporation occurs during melting, adjust the volume to 150 ml with deionized water.

5. Cool the agarose solution to approximately 60°C and add 5 µl of ethid-

ium bromide stock solution. Slowly pour the agarose into the gel-casting tray. Remove any air bubbles by trapping them in a 10 ml pipette.

6. Position the comb approximately 1.5 cm from the edge of the gel. Let the agarose solidify for approximately 30–60 minutes. After the agarose has solidified, remove the comb with a gentle back and forth motion, taking care not to tear the gel.

7. Remove the tape from the ends of the gel-casting tray and place the tray on the central supporting platform of the gel box. For safety purposes, the electrophoresis apparatus should be always placed on the laboratory bench with the positive electrode (red) facing away from the investigator, that is away from the edge of the bench.

8. Add electrophoresis buffer to the buffer chamber until it reaches a level of 0.5–1 cm above the surface of the gel.

9. Load the samples into the wells using a yellow tip. Place the tip **under** the surface of the electrophoresis buffer just **above** the well. Expel the sample slowly, allowing it to sink to the bottom of the well. Take care not to spill the sample into a neighboring well. During sample loading, it is very important to avoid placing the end of the tip into the sample well or touching the edge of the well with the tip. This can damage the well, resulting in uneven or smeared bands. **Note:** samples must be loaded in sequential sample wells. When loading fewer samples than the number of wells it is preferable to leave the wells nearest the edge of the gel empty.

10. First load 8 µl of the 1 kb ladder standard DNA. Next load the entire sample (35 µl) using a P200 Pipetman. Load the samples in the following order: 1H, 2H, 3H, and 4H.

11. Place the lid on the gel box and connect the electrodes. DNA will travel towards the positive (red) electrode positioned away from the edge of the laboratory bench. Turn on the power supply. Adjust the voltage to approximately $1\,\text{V}\,\text{cm}^{-1}$. For example, if the distance between electrodes (not the gel length) is 40 cm the voltage should be set to 40 V in order to obtain a field strength of $1\,\text{V}\,\text{cm}^{-1}$.

12. Continue electrophoresis until the tracking dye moves at least two-thirds of the gel length. It will take the tracking dye approximately 17 hours to reach this position on a gel 20 cm long.

Next day

1. Turn the power supply off and disconnect the positive (red) lead from the power supply. Remove the gel from the electrophoresis chamber. To avoid electric shock always disconnect the red (positive) lead first.

2. Wrap the gel-casting tray with saran wrap and store in a 4°C refrigerator. Gels can be stored this way for two to four days.

References

Adamson, A.W. (1976) *Physical Chemistry of Surfaces*. John Wiley & Sons Publishers, New York.

Hiemenz, P.C. (1977) *Principles of Colloid and Surface Chemistry*. Marcel Decker Inc., New York.

Perbal, B. (1988) *A Practical Guide to Molecular Cloning*, 2nd edn. John Wiley & Sons Publishers, New York, Chichester, Brisbane, Toronto, and Singapore.

Sealey, P.G. and Southern, E.M. (1982) Gel electrophoresis of DNA. In *Gel Electrophoresis of Nucleic Acids: A Practical Approach*, D. Rickwood and B.D. Hames (eds). IRL Press, Oxford.

Sharp, P.A., Sugden, B., and Sambrook, J. (1973) Detection of two restriction endonuclease activities in *Haemophilius parainfluenzae* using analytical agarose – ethidium bromide electrophoresis. *Biochemistry*, **12**, 3055–60.

Surzycki, S.J. (2000) *Basic Techniques in Molecular Biology*. Springer-Verlag, Berlin, Heidelberg, and New York.

SECOND LABORATORY PERIOD

Experiment 3: Southern blotting

Introduction

You will continue multi-locus analysis of your DNA by transferring DNA from the gel prepared in experiment 2 to the nylon membrane. This process is called Southern blotting. The DNA fragments are transferred to the membrane at the same positions as they are on the gel. The blotting technique that we will be using relies on a flow by capillary action of a neutral transfer solution from a reservoir through the gel to a membrane overlaid by a stack of dry paper towels. Prior to transfer, DNA fragments will be denatured *in situ* to single-stranded DNA that can be bound to the membrane and hybridized to probe. DNA fragments will be immobilized to the membrane by UV light irradiation.

Background

Southern (1975) introduced immobilizing target DNA to a membrane for hybridization studies. The technique permits hybridization of various probes to immobilized target DNA under controlled conditions. The Southern blot method, originally described by Southern (1975), combines the high resolving power of agarose gel electrophoresis in the separation of DNA fragments with the specificity of DNA–DNA hybridization reactions. The basic principle of the technique is that DNA fragments, which are separated by agarose gel electrophoresis, are transferred and immobilized to a solid support, such as a nylon membrane. Once immobilized, the DNA is available for hybridization with labeled DNA or RNA probes. This technique is applicable to the analysis of small, cloned DNA fragments, as well as to the analysis of genomic DNA. DNA transfer to solid support is generally accomplished by capillary methods, but electroblotting, positive pressure, and vacuum transfer procedures can also be used (Peferoen et al., 1982; Smith et al., 1984). These other methods are in general faster than capillary transfer, but are less efficient and require expensive equipment.

There are two capillary transfer methods: upward capillary transfer and downward capillary transfer. Upward capillary or "standard" transfer results in very efficient transfer of DNA or RNA of all sizes but requires overnight exposure. Downward capillary transfer is just as efficient as upward transfer and requires a much shorter transfer time (3–4 hours). Capillary transfer can be carried out with neutral or alkaline transfer solutions (Chomczynski, 1992).

Alkaline transfer of DNA is only possible with a positively charged nylon membrane. Capillary alkaline transfer is faster because the neutralization step is omitted from the procedure and the time required for transfer is shorter. An alkaline transfer solution can cause some depurination of the DNA and frequently results in weaker hybridization signals than the neutral transfer. Alkaline blotting is not recommended for genomic DNA transfers or when reprobing of the membrane with another probe is planned.

Safety precautions

The agarose gel contains ethidium bromide, which is a mutagen and suspected carcinogen. Students should wear gloves when handling these gels. Powder free gloves only should be used because the procedure described uses chemiluminescent for the detection of hybridization. The presence of talcum powder will result in the formation of a "spotted" background. Discard the used gel into the designated container.

When viewing and photographing the gel with a UV transilluminator, gloves, UV-protective glasses, and a facemask should be used all times.

Technical tips

Different types of support membranes can be used for DNA fingerprinting experiments. However, the use of chemiluminescent detection requires a positively charged nylon membrane. These membranes have a very low background with chemiluminescent detection. Some producers have developed a chemically optimized, positively charged nylon membrane for chemiluminescent detection. Evaluation of all types of commercially available positively charged nylon membranes showed that the best membrane for chemiluminescent detection is a Magna Graph membrane, which is manufactured by Osmonics/MSI Co. (Surzycki, 2000).

The best signal to noise ratio for a Magna Graph membrane is achieved when DNA is cross-linked to a membrane by UV light irradiation.

Protocol

1. Transfer the gel to a glass Pyrex dish and trim away any unused areas of the gel with a scalpel. Cut off the gel below the bromophenol blue dye. This part of the gel does not contain DNA fragments. Cut the lower corner of the gel at the bottom of the lane with size standards. This will provide a mark with which to orient the hybridized bands on the membrane with the bands in the gel. **Note:** because the gel is thinner in the well area, the transfer solution may pass preferentially through this part of the gel causing uneven DNA transfer. Remove this part of the gel.

2. Transfer the gel to a UV transilluminator. Place an acetate sheet on top of the gel and draw an outline of the gel with a felt marker pen. Mark the

positions of the wells and the position of the cut corner. **It is very important that your drawing be as precise as possible.** Label the contents of each well on the acetate sheet and mark the bottom left corner of the gel (under well number 1). This will help you locate the positions of the hybridization signals in your Southern blots. Turn on the transilluminator and mark the positions of standard DNA bands on the acetate sheet.

3. Photograph the gel. Use a setting of 1 second at F8 when using a Polaroid 667 film pack. One can also use a computer-imaging system for recording the results.

4. Transfer the gel back to the Pyrex dish and add enough 0.25 N HCl to allow the gel to move freely in the solution. This will take approximately 150–200 ml of solution for a standard gel size.

5. Place the dish on an orbital shaker and incubate for 10 minutes rotating at 10–20 r.p.m. **Note:** this procedure breaks large DNA molecules by depurination, thereby facilitating their efficient transfer onto the membrane. It is important not to let the hydrolysis reaction proceed too far, otherwise the DNA is cleaved into fragments that are too short to bind efficiently to the membrane (less then 200 bp).

6. Decant the acid carefully holding the gel with the palm of your gloved hand.

7. Rinse the gel for 10–20 seconds in 200 ml of distilled water. Discard the water and proceed immediately to the next step.

8. Add 200 ml of denaturation solution to the dish and gently agitate it for 20 minutes on a rotary shaker.

9. Decant the denaturation solution as described above and repeat step 8 one more time.

10. Add 200 ml of water and rinse the gel for 10–20 seconds in order to remove most of the denaturation solution trapped on the surface of the gel. Decant the water, holding the gel with the palm of your gloved hand. Be very careful during this procedure because the denaturation solution contains NaOH making the gel very slippery.

11. Add 100–200 ml of neutralization solution to the dish and treat the gel for 20 minutes with gentle agitation.

12. Discard the neutralization solution and repeat step 11 one more time.

13. While the gel is being treated, prepare the nylon membrane for transfer. Cut the nylon membrane to the size of the gel. Use the outline of the gel drawn on the acetate sheet as a guide. Use gloves and only touch the edges of the nylon membrane. Place the membrane in a separate Pyrex dish filled with distilled water. Leave the membrane in water for 1–2 minutes. Decant the water and immerse the membrane in ten times SSC. Cut three sheets of Whatman 3 MM paper to the size of the nylon membrane.

14. Prepare a long strip of Whatman 3 MM paper to use as a wick. The wick should be approximately 30 cm long and 10 cm wide.

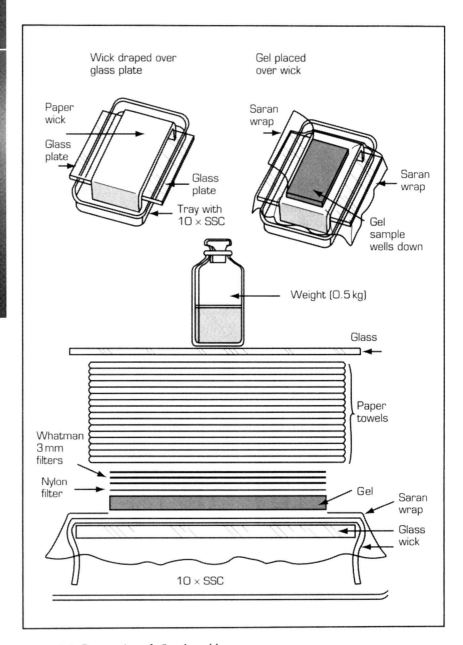

Figure 2.3 Preparation of a Southern blot.

15. Assemble the blot sandwich. Refer to Fig. 2.3 for illustration of the assembly. Add 400–600 ml of ten times SSC to a large Pyrex dish. Place a glass platform across the center of the dish and cover it with the wick. Make sure that both ends of the wick are immersed in SSC solution. Wet the wick with 10–20 ml of ten times SSC and remove trapped air bubbles by rolling a 10 ml glass pipette over it. Carefully lift the gel from the Pyrex dish and place it in the center of the wick with the **sample wells down**. Smooth the gel and remove trapped air bubbles by gently rolling a glass pipette over the surface. **Note:** the gel is now upside-down with the well openings facing the wick. This is necessary in order to obtain the best results during transfer (sharper resolution due to less diffusion during the transfer) and to maintain the left-to-right sample orientation on the membrane.

16. Cover the entire dish, including the surface of the gel, with saran wrap. With a razor blade "cut away" the saran wrap covering the gel itself and discard it. This will leave an opening over the gel while the remaining area of the wick will be covered by saran wrap.
17. Place the nylon membrane on top of the gel. Add 5 ml of ten times SSC to the top of the membrane and remove air bubbles by rolling a pipette over it. Cut the left bottom corner of the membrane to coincide with the cut made in the gel.
18. Place three sheets of dry Whatman 3 MM paper, prepared in step 13, on the top of the membrane. Place several inches of paper towels on top of the 3 MM paper. Because the wick area is protected by saran wrap, it is not necessary to cut the paper towels to the size of the gel. Place a glass plate on top of the paper towel stack and weigh it down with a 1 l Erlenmeyer flask filled with 500 ml of water. Allow a minimum of 17 hours for the transfer. **Note:** to prevent the gel from collapsing, the weight placed on the top of the stack should never exceed 500 g (i.e. approximately 500 ml of water).

Next day

1. Disassemble the blot. Remove the weight, glass plate, and paper towels. Using forceps remove the membrane and place it "DNA side" up (the side that was in contact with the gel) on a clean sheet of Whatman 3 MM paper. Mark the DNA side with pencil on the corner of the membrane.
2. Place the membrane on a sheet of dry Whatman 3 MM paper. Do not allow the membrane to dry at any time. Place the membrane into a UV oven.
3. UV irradiate the damp membrane to cross-link DNA to the membrane using the automatic setting of the UV oven. Irradiate **both sides** of the membrane. Alternatively, you can wrap the membrane in aluminum foil and bake it in an oven at 80°C for 1 hour. The baking step immobilizes DNA on the membrane.
4. Place the membrane into a plastic bag and store it at room temperature.

References

Chomczynski, P. (1992) One-hour downward alkaline capillary transfer for blotting of DNA and RNA. *Anal. Biochem.*, **201**, 134–9.

Peferoen, M., Huybrecht, R., and De Loof, A. (1982) Vacuum-blotting: a new simple and efficient transfer of proteins from sodium dodecyl sulfate–polyacrylamide gels to nitrocellulose. *FEBS Lett.*, **145**, 369–72.

Smith, M.R., Devine, C.S., Cohn, S.M., and Lieberman, M.W. (1984) Quantitative electrophoretic transfer of DNA from polyacrylamide or agarose gels to nitrocellulose. *Anal. Biochem.*, **137**, 120–4.

Southern, E.M. (1975) Detection of specific sequences among DNA fragments separated by gel electrophoresis. *J. Mol. Biol.*, **98**, 502–17.

Surzycki, S.J. (2000) *Basic Techniques in Molecular Biology*. Springer-Verlag, Berlin, Heidelberg, and New York.

THIRD LABORATORY PERIOD

In this laboratory period you will carry out two procedures. First, you will prepare probe for hybridization using a PCR. Second, you will hybridize the probe to membrane containing *Hae*III-restricted DNA prepared previously.

Experiment 4: preparation of probe and hybridization

Introduction

The hybridization probe will be labeled with digoxigenin (DIG) using dUTP-DIG as labeled substrate. In this substrate dUTP is linked via a spacer arm to the steroid hapten DIG. In this reaction, the ratio of dUTP-DIG to dTTP will be 1:2, assuring high density labeling of the probe. You will amplify tandem repeats present in the RF form of bacteriophage M13 genome using specific primers that span the tandem repeat region. An amplified tandem repeat with core sequence GAGGGTGGNGGNTCT is complementary to a large number of hypervariable mini-satellites present in the human genome.

Hybridization will be carried out using a hybridization oven in specially designed hybridization roller bottles. The hybridization procedure described uses low stringency conditions for hybridization and washing. These conditions of hybridization make it possible to hybridize probe to a family of hypervariable mini-satellites that are closely related to the core sequence.

Background

DNA probe preparation

Preparation of probes for hybridization involves *in vitro* incorporation of reporter molecules into nucleic acids. These reporters can be incorporated at one or both ends of nucleic acid molecules, giving specific, low-density labeled probes. High-density labeling is usually achieved by incorporating the reporters uniformly throughout the entire length of nucleic acid molecules. For hybridization work, internally labeled probes are preferred since they provide the strongest hybridization signal.

Two types of reporter molecules are presently in use, radioactive reporters and non-radioactive reporters.

Radioactive reporters are usually tagged with ^{32}P or ^{35}S isotopes and can be directly detected using X-ray film in a process called autoradiography.

Non-radioactive reporters can be fluorescent tags, permitting their direct detection or specific ligands such as biotin, haptens, and hapten-like molecules, the detection of which is indirect.

Two methods are used for detecting non-radioactive reporters: direct and indirect. The direct method uses reporter enzyme conjugation to the probe. Probes obtained by these methods have lower levels of background, thus allowing longer exposure time and, consequently, better sensitivity. However, the presence of enzyme (conjugated to the probe) during hybridization requires modification of the standard hybridization protocol in order to protect the enzyme from inactivation. We will use this method for single-locus analysis.

Indirect detection is carried out using enzymatic reactions catalyzed by either horseradish peroxidase or alkaline phosphatase. These enzymes can be conjugated directly to a secondary molecule that has a very high affinity for a specific ligand (for example biotin–avidin complexes) or they can be conjugated to antibody against a hapten tag. Since enzymes are only used for the detection of labeled hybrids and are not present during hybridization, standard hybridization protocols can be used. This method will be used in multi-locus analysis.

Two types of substrates are used for detecting enzyme activity at the site of hybridization: colorimetric substrate and chemiluminescent substrate.

Colorimetric detection uses a soluble, colorless substrate that is converted into an insoluble, colored product precipitated directly on the membrane. We will use this method for determining the concentration of DIG label in our probe.

Chemiluminescent detection uses a chemiluminescent substrate that is converted by enzyme into a light-emitting substance easily detected by standard photographic film. The sensitivity of the chemiluminescent method is much greater than the colorimetric method and even exceeds the sensitivity of radioactive labeling (Höltke et al., 1990; Kessler et al., 1990). The chemiluminescent substrates now in use allow the detection of 0.03 pg or less DNA on the membrane in approximately 1–2 hours. The same amount of DNA would require 17–20 hours of exposure to detect a ^{32}P-labeled probe. We will use this substrate for detecting hybridization in multi-locus analysis.

Preparation of uniformly labeled probes, independent of the nature of the reporter, is carried out using four basic procedures: nick translation, random priming, RNA probe synthesis, and PCR.

Nick translation

In this method, labeled DNA is synthesized using *Escherichia coli* DNA polymerase. Single-stranded nicks are introduced at random in the double-stranded DNA template with the DNA exonuclease DNase I. DNA polymerase I initiates DNA synthesis at the 3′-end of the nick, whereas at the 5′-end of the nick the 5′ to 3′ exonuclease activity of this enzyme excises nucleotides. The position of the nick is "translated" downstream, with labeled nucleotides replacing non-labeled nucleotides between the original site of the nick and the new position of the nick. The resulting probe is labeled on both strands with labeled (newly synthesized) DNA interspersed with unlabeled (original template) DNA. The method works best with linear DNA fragments that are larger than 500 bp. The specific activity of these probes is somewhat lower, as compared to other methods, but nick-translated probes generally give very strong hybridization signals (Davis et al., 1994).

Random priming

In this method DNA polymerase synthesizes a newly labeled DNA by template-dependent extensions of random hexamer primers (Feinberg and Vogelstein, 1983). Polymerases lacking 5′ to 3′ exonuclease activity, such as the Klenow fragment of *E. coli* DNA polymerase I, are used in this reaction. The template strand remains unlabeled whereas the newly synthesized strand is completely labeled. The enzyme can initiate and synthesize several new strands from every template, resulting in net synthesis of large amounts of labeled product, many times exceeding the amount of template input. The specific activity of probe produced by random priming is very high. The method is relatively insensitive to the purity of the DNA template or its size and creates probes of 100–600 bp long. Random priming is particularly well suited for the preparation of non-radioactive probes of very high specific activity.

PCR probes

Two alternative techniques are used for preparing probes using PCR. In the first procedure, uniformly labeled probes are generated by incorporation of a tagged nucleotide during PCR (Kessler, 1992; McCreery and Helentjaris, 1994; Yamaguchi at al., 1994). In the second approach, a large number of specific DNA target molecules are synthesized by PCR, which are subsequently used as a template for preparing random-primer or nick-translated probes (Rost, 1995). PCR labeling can be done using either a genomic template without cloning the DNA fragment in question or from a DNA fragment cloned into plasmid. Specific primers are required for the former but not for the latter PCR procedure. The advantages of generating DNA probes by PCR are numerous. First, large amounts of probe with high label density can be synthesized from very little DNA. Second, probes can be prepared using either purified or partially purified DNA as the source of the template. Third, preparation of the probe is highly flexible and does not depend on restriction enzyme site location. Fourth, it is possible to prepare specific, single-stranded probes using a single primer. The main disadvantage is the difficulty in amplifying most sequences larger than 3,000 bp in length. We will prepare multi-locus probe using the PCR method.

DNA hybridization

The hybridization reaction is the formation of partial or complete double-stranded nucleic acid molecules by sequence-specific interaction of two complementary single-stranded nucleic acids. The hybridization reaction, using labeled probes, is the only practical way of detecting the presence of specific nucleic acid sequences in a complex nucleic acid mixture. The most frequently used hybridization technique is the membrane hybridization technique. Hybrid formation between complementary strands is commonly called a re-association, renaturation, or re-annealing reaction. The reverse reaction is called strand separation, dissociation, or melting of the DNA.

Kinetics of the hybridization reaction

The hybridization reaction proceeds in two steps: the nucleation reaction and "zippering" reaction. Nucleation is the formation of short hybrids of a few bases long between reacting strands. The nucleation reaction, with some approximation, is a diffusion-limited reaction defined by the Smoluchowski and Deby equation. Therefore, the reaction rate depends on the solvent viscosity, temperature, and ionic strength of the medium (Wetmur and Davidson, 1968; Chang et al., 1974).

Many nucleation events will take place until by chance the correct base pair is formed. A rapid zippering process follows this. The zippering reaction is an extension of the hybrid from the nucleation site throughout the entire molecule. This reaction is very fast and largely independent of the factors mentioned above. Thus, the limiting step in a hybridization reaction is the nucleation reaction and not the zippering process. The overall hybridization rate is dependent on a nucleation rate constant (k_n), the probe length, and target complexity as described by the equation

$$k_2 = \frac{k_n}{\sqrt{L}} \tag{2.4}$$

where k_n is the nucleation rate constant and L is the probe length in bases. The nucleation rate k_n for monovalent cation concentrations commonly used in filter hybridization (0.2–4.0 M) and solvent viscosity comparable to 1.0 M NaCl can be evaluated using the equation (Orosz and Wetmur, 1977)

$$k_n = (4.35\log[M^+] + 3.5)10^5 \tag{2.5}$$

Temperature of the hybridization

In order to determine the temperature of hybridization, the melting temperature (T_m), which is defined as the temperature at which the DNA is 50 percent denatured, must be known. Melting of DNA is an intramolecular, first-order reaction and, therefore, is independent of substrate concentration. It depends only on the base composition of the duplex and the composition of the solvent. Equation (2.6), which was first empirically established by Marmur and Doty (1962) for molecules shorter than 500 bp, describes the T_m for DNA:DNA duplexes:

$$T_m = 81.5 + 0.41(\%GC) + 16.6 \log[M^+] \qquad (2.6)$$

where [M^+] is the molar concentration of the monovalent cations Na$^+$ or K$^+$ and %GC is the percent of GC bases in hybrid DNA. This equation was modified to incorporate DNA:RNA and RNA:RNA hybrids and extends the monovalent cation concentration range from 0.01 to 4.0 M, the concentrations at which most hybridization reactions are carried out (for a detailed discussion of modified equations see Wetmur (1991) and Surzycki (2000)).

It is obvious that hybridization should be carried out at temperatures below the T_m temperature calculated from equation (2.6). The difference between the T_m temperature and hybridization temperature is defined as the "criterion of hybridization" (Britten et al., 1974). The overall rate of the hybridization reaction k_2 is strongly affected by temperature. This dependence has a bell-shaped curve and increases as the criterion is increased, reaching a broad maximum between 20 and 25°C below the T_m for the DNA:DNA hybrid. As the temperature of hybridization falls further below the T_m, the hybridization rate decreases very fast due to intramolecular base pair formation that decreases the availability of nucleation sites. Therefore, the optimal temperature for hybrid formation (T_h) for a DNA:DNA hybrid is

$$T_h = T_m - (20-25°C) \qquad (2.7)$$

At T_h temperature, not only does hybrid formation occur at maximum speed, but it is most perfect. This is because at T_m the formation of perfectly matched hybrids by the zippering reaction is faster than the formation of mismatched hybrids (10 percent or more mismatches). Moreover, the T_m of imperfectly matched hybrids is lower than perfectly matched hybrids (approximately 1°C for each percent of mismatch). Consequently, the maximum hybridization rate of mismatched hybrids occurs much below the hybridization temperature of a perfect hybrid. The formation of poorly matched hybrids shows a similar bell-shaped dependence on temperature, but the maximum rate (k_2) is several orders slower than for a well-matched hybrid and the entire curve is displaced towards lower temperatures.

Hybridization time

Membrane blot hybridization is usually performed with a large excess of probe DNA. For example, approximately 5–10 µg of eucaryotic genome is used in Southern blot analysis. Assuming that a single gene size is 2,000 bp, the genome size is 3.15×10^9 bp (e.g. human genome), and only 2 percent of the DNA bound to the nylon membrane is open for hybridization (Vernier et al., 1996), the amount of a single gene present on a Southern blot is approximately 0.06×10^{-6} µg. The concentration of probe is usually 20–25 ng ml^{-1}. If the probe is the same size as the genomic target and 10 ml of hybridization solution is used, the amount of probe present is 0.2–0.25 µg or a nearly 3×10^6-fold excess of probe DNA over target DNA. Using these hybridization conditions, the reaction is pseudo first order and its half time is

$$t_{1/2} = 2/k_2 C_0 \qquad (2.8)$$

where k_2 is rate of hybridization as calculated from equation (2.4) and C_0 is the initial concentration of probe in moles of nucleotides per liter (Wetmur, 1991, 1995). Since most hybridization reactions are carried out at 1 M Na$^+$ the k_n for this reaction is equal to $3.5\ 10^5$ M^{-1}s^{-1} (from equation 2.5). The concentration of the probe used in Southern hybridization is usually equal to $C_0 = 6.1\ 10^{-8}$ M (20 ng ml^{-1}) and k_2 is $7.8\ 10^3$ M^{-1}s^{-1} for a single gene probe calculated from equation (2.4). Using equation (2.8), the $t_{1/2}$ of the reaction can be calculated to be approximately 1 hour.

If an excess of double-stranded probe is hybridized to an immobilized target, self-annealing of the probe limits the time of effective hybridization to approximately two to three times the $t_{1/2}$ calculated for a single-stranded probe. Thus, the above reaction will reach completion in approximately 2–3 hours due to self-annealing of the probe. The half-time for pseudo first-order hybridization can be approximated (in hours) when double-stranded probe is used with standard conditions of hybridization (i.e. 1 M salt at T_h equal to 20–25°C below T_m) from the equation (Sambrook et al., 1989)

$$t_{1/2} = \frac{1}{X} \times \frac{Y}{5} \times \frac{Z}{10} \times 2 \qquad (2.9)$$

where X is micrograms of probe added, Y is the probe complexity in kilobases (length of probe), and Z is the volume of hybridization in milliliters. For example, for the DNA hybridization reaction described above, that is hybridization with a 2,000 bp probe at a concentration of 0.02 µg ml^{-1} in

10 ml, the $t_{1/2}$ calculated from equation (2.9) is 2 hours, a value close to that calculated from equation (2.8).

However, hybridization reactions are usually carried out for 13–17 hours in order to increase the signal strength. This is because probes can form extended networks in which the single-stranded tails from one duplex hybridize to a complementary single-stranded tail of another duplex. This network formation occurs five to six times slower than the annealing of a single-stranded probe and its formation will increase the hybridization signal intensity (Geoffrey et al., 1987).

The hybridization kinetics described above only apply strictly to liquid hybridization. When nucleic acid is immobilized on a membrane, the hybridization rate is decreased because the rate of access of probe to target is decreased significantly. Consequently, the nucleation rate (k_n) is two to four times lower than the nucleation rate in liquid hybridization, but the effect of T_m and ionic strength are not changed.

Hybridization reaction solution

Because the rate of the nucleation reaction strongly depends on salt concentration, hybridization should not be carried out at low salt concentration. In order to achieve fast hybridization rates and, consequently, to form the most perfect hybrid, a monovalent cation concentration between 0.75 and 1.0 M is used. Because DNA used in most hybridization reactions has approximately 50 percent GC, the calculated T_h is usually between 72 and 78°C, as calculated from equation (2.7). Single-stranded DNA (probe) is particularly prone to depurination at temperatures higher than 50°C. Prolonged hybridization at high temperature will result in degradation of probe and target DNAs (Blake, 1995). In order to lower the hybridization temperature, the hybridization reactions are usually carried out in the presence of denaturing solvents while maintaining high ionic strength (Hutton, 1977). The most commonly used solvent in membrane hybridization is formamide. This solvent lowers, on average, the T_m of DNA by 0.7°C per 1 percent formamide. The effect of formamide is greater on AT nucleotide pairs than on GC pairs (Anderson and Young, 1985). Formamide has no apparent effect on the rate of hybridization at concentrations between 30 and 50 percent for membrane hybridization, making it an ideal solvent for lowering the incubation temperature. The concentration of formamide most frequently used is 50 percent. This lowers the temperature required for hybridization by approximately 32°C, thereby allowing hybridization of most DNA at temperatures below 50°C without substantially lowering the rate of hybridization.

Because of the high toxicity of formamide, a new non-toxic compound was recently introduced. This is Dig Easy Hyb solution, which is manufactured by Roche Molecular Biochemicals. The temperature of hybridization in this solution should be calculated with the same equations that are used for hybridization in 50 percent formamide and 1 M Na⁺. Thus, the equation for T_m simplifies to

$$T_{m(\text{Dig Easy})} = 49.82 + 0.41(\%GC) \tag{2.10}$$

We will use Dig Easy Hyb solution in our hybridization experiment.

Washing reaction

The hybridization reaction is followed by a washing reaction that removes any unhybridized probe and melts mismatched hybrids. This reaction, unlike the hybridization reaction, is a first-order reaction and depends on the thermal stability of the hybrid (T_m). The T_m of the hybrid is lowered by approximately 1°C for each percent of mismatch (Bonner et al., 1973). In order to obtain 95 percent faithful hybrids the washing reaction is carried out at a washing temperature expressed by the equation

$$T_w = T_m - 5°C \tag{2.11}$$

where T_m is melting temperature calculated from equation (2.6) or (2.10). Reactions carried out at this temperature are called high stringency washes. Reactions performed at temperatures lower than this are usually referred to as "low stringency" washing reactions.

Technical tips

Hybridization can be performed in roller bottles, glass or plastic dishes, or in sealed plastic bags. The hybridization and subsequent procedure for signal detection described here uses a hybridization oven and large roller bottles. This is the most convenient method to use in a large-class situation. No more than one membrane should be placed into the roller bottle. Some membrane overlapping will not affect the hybridization results.

The volume of the reagents used in each step will be substantially increased by using other containers for hybridization and subsequent membrane treatment. The minimum volume of the hybridization probe solution should be 0.2 ml solution per 1 cm² of membrane surface area. The volume of all other solutions should be approximately four to five times larger than for hybridization. Care should be taken that the membrane is sufficiently covered with solutions at all times and that it can float freely in the container.

The concentration of DIG-labeled probe should be 10–15 ng DIG label per milliliter. Increasing the probe concentration will not increase the hybridization signal, but will substantially increase the background. Careful determination of the DIG concentration in the probe is crucial for the success of this experiment.

One of the great advantages of the DIG system is the stability of the labeled probe. The probe can be stored practically for an indefinite time at −20°C. Moreover, the hybridization solution contains unannealed DIG-labeled probe and can be reused several times. This solution can be stored for at least one to two years at −20°C.

The hybridization membrane prepared in this experiment will be used in experiment 3. The membrane should be stored after exposure to X-ray film in the bag with chemiluminescent substrate at 4°C. If longer storage is required (e.g. several months) the chemiluminescent substrate should be removed from the bag and replaced with a solution containing two times SSC and 0.1 percent SDS (sodium deodecyl sulfate).

Protocol

Start this laboratory with pre-hybridization as described in steps 1–3 of the pre-hybridization and hybridization procedure. Pre-hybridization should be carried out for 2–3 hours. During this time, prepare and determine the concentration of probe as described in the following protocol.

Probe preparation and determination of digoxigenin label concentration

1. Place a sterilized 1.5 ml microfuge tube on ice and add the ingredients following Table 2.2. First, calculate the amount of water necessary for the desired volume and add it to the tube. Then add buffer and the remaining components. Add enzyme last and mix by pipetting up and down several times. Never mix by vortexing. Taq polymerase is **very sensitive to vortexing**.

2. Load the reaction into 20 μl capillary tubes. Insert the open end of the capillary tube into the white silicon tip of the micro-dispenser approx-

Table 2.2 Preparation of DIG-labeled probe

Ingredient	Add	Final concentration
Buffer high Mg^{2+} (ten times)	2.0 μl	one times
4dNTP labeled (ten times)	2.0 μl	200.0 mM
Taq polymerase (1–5 u μl^{-1})	1.0 μl	1–5.0 u
Primer M13 V F (5 μM)	2.0 μl	5.0 pM
Primer M13 V R (5 μM)	2.0 μl	5.0 pM
DNA RF M13 (2 ng μl^{-1})	2.0 μl	0.4 ng
Water	9.0 μl	
Total	20.0 μl	

imately 5 mm deep and draw the reaction mixture into it by slowly turning the micro-dispenser knob counterclockwise. Position liquid in the middle of the capillary and seal it by flaming the ends. Only a few seconds of heating the extreme tip of the capillary is necessary. Check the seal by gently turning the micro-dispenser knob back and forth. If the liquid does not move in the tube, the tube is sealed. Flame seal the other end of a capillary tube. Place the sealed tube into a microfuge tube **with the end sealed last on top**.

3. Set the cycling conditions as follows: D (denaturation) = 94°C for 15 seconds, A (annealing) = 55°C for 10 seconds, and E (elongation) = 72°C for 60 seconds. Start with 94°C for 2 minutes. End with 72°C for 5 minutes. The total time of the PCR will be approximately 50 minutes.

4. Place the capillary tubes into the DNA thermal cycler inserting the end sealed first into the holder. Start the machine and observe the end of the capillary tube extruding from the holder carefully. If the second end is improperly sealed, liquid will rise to the top of the capillary tube when the temperature reaches approximately 90°C. If this happens, stop the cycler and reseal the open end. Restart the cycling again.

5. After cycling is complete, remove the liquid from the capillary tube into an appropriately labeled 1.5 ml centrifuge tube. Holding the capillary tube horizontally, gently snap off one end. Insert the open end of the capillary tube into the white silicon tip of the micro-dispenser, approximately 5 mm deep. Snap off the other end of the capillary tube. Remove the amplified sample from the capillary tube by slowly turning the micro-dispenser knob clockwise.

6. Add 9 µl of water to ten microfuge tubes. Prepare a tenfold serial dilution of the newly labeled DNA. Serially dilute as follows: 1:10, 1:100, 1:1,000, 1:10,000, and 1:100,000. Take 1 µl of the labeled probe and add it to the first dilution tube (S1). Mix well by pipetting up and down. Withdraw 1 µl from this tube and add it to the second dilution tube (S2). Mix well by pipetting up and down. Using the same procedure, prepare 1:1,000 (S3), 1:10,000 (S4), and 1:100,000 (S5) dilutions. Use the same yellow tip for all serial dilutions. Using the remaining five tubes, prepare a similar serial dilution for the DIG-labeled standard (5 ng µl^{-1} DIG-labeled plasmid).

7. Cut a piece of nylon membrane 7 cm × 2 cm and place it at the bottom of a Petri dish. Spot 1 µl of each DIG-labeled dilution onto the membrane. Start from the largest dilution (1:100,000) and continue spotting towards the less dilute samples. Use the same yellow tip for spotting the entire series. Repeat this procedure using standard DNA. Place each standard directly under the appropriate dilution of your labeled DNA. Arrange the spots on the membrane in the following way:

Sample DNA	1:100,000	1:10,000	1:1,000	1:100	1:10
Standard DNA	1:100,000	1:10,000	1:1,000	1:100	1:10

8. Let the spotted samples dry completely under a heat lamp for 2–4 minutes. Wet the membrane with ten times SSC by placing it onto wet

Whatman 3 MM filter paper. Transfer the membrane to dry Whatman 3 MM filter paper and immobilize the DNA on the damp membrane by UV cross-linking in a UV cross-linking oven. The membrane must be damp in order to cross-link the DNA. Best results are obtained by application of a calibrated UV light source such as a Stratalinker UV oven. Use the "autocross-linking" function of the UV oven.

9. Transfer the membrane to a Petri dish and add 15 ml of buffer A. Incubate on a rotary shaker at 10–20 r.p.m. for 5 minutes. Discard buffer A.
10. Add 15 ml of buffer B and incubate on a rotary shaker as above for 5 minutes. Discard buffer B.
11. During incubation in buffer B, prepare anti-DIG-AP solution. Add 10 ml of buffer B and 2 µl of anti-DIG-AP conjugate antibody to a sterile 15 ml plastic conical centrifuge tube. Mix well by inverting the tube several times.
12. Add anti-DIG-AP solution to the membrane and incubate with gentle rotation for 5 minutes at room temperature. Discard the antibody solution.
13. Add 15 ml of buffer A to the membrane and wash with gentle rotation for 5 minutes. Discard buffer A and repeat the wash once more.
14. Add 15 ml of buffer C and incubate on the rotary shaker for 2 minutes. Discard buffer C.
15. Prepare color development solution as follows. Add 10 ml of buffer C to a 15 ml plastic conical centrifuge tube. Add 45 µl of NBT solution and 35 µl of X-phosphate solution to the tube. Close the tube and mix by inverting several times. Protect the solution from direct light.
16. Add color development solution to the membrane. Make sure the solution covers the membrane. Cover the dish with aluminum foil to protect it from light. Incubate **without shaking** for 30–60 minutes checking occasionally for color development.
17. Compare the spot intensities of probe with control DNA and estimate the concentration of DIG-labeled probe.

Pre-hybridization and hybridization procedure

1. Place the dry membrane into a roller bottle. Make sure that the side of the membrane with DNA is facing away from the glass.
2. Pour 10 ml of Dig Easy Hyb solution into the roller bottle. Close the bottle tightly and label it with your group number.
3. Place the roller bottle into the hybridization oven and allow it to rotate at a **slow speed** (2–4 r.p.m.) for 1–3 hours at 34°C.
4. Ten minutes before the end of pre-hybridization begin to prepare the probe for hybridization. Add 10 ml of Dig Easy Hyb solution to a 15 ml conical centrifuge tube. Add DIG-labeled probe to the Dig Easy Hyb to a final concentration of 15–20 ng ml^{-1}. Close the tube tightly. Place the tube into an 80°C water bath and incubate for 10 minutes to denature the double-stranded DNA probe.

5. Retrieve the roller bottle from the hybridization incubator. Open the bottle and pour the pre-hybridization solution into a storage bottle. Pre-hybridization solution can be stored and used again.

6. Remove the tube with the probe from the 80°C water bath and add the labeled probe to the hybridization tube. Return the roller bottle to the oven and allow it to **rotate slowly** at 34°C overnight.

Next day

1. The next day, remove your roller bottle from the hybridization oven and pour off the hybridization solution into a 15 ml centrifuge tube. The probe can be stored in a –20°C freezer and reused several times.

2. Add 20 ml of washing solution II (two times SSC and 0.1 percent SDS) to the roller bottle. Place the bottle into the hybridization oven and allow it to rotate at slow speed until the next laboratory period.

References

Anderson, M.L.M. and Young, B.D. (1985) Quantitative filter hybridization. In *Nucleic Acid Hybridization: A Practical Approach*, B.D. Hames and S.J. Higgins (eds), pp. 73–111. IRL Press, Oxford.

Blake, R.D. (1995) Denaturation of DNA. In *Molecular Biology and Biotechnology. A Comprehensive Desk Reference*, R.A. Meyers (ed.), pp. 207–10. VCH Publishers, Inc., New York.

Bonner, T.I., Brenner, D.J., Neufeld, B.R., and Britten, R.J. (1973) Reduction in the rate of DNA reassociation by sequence divergence. *J. Mol. Biol.*, **81**, 123–35.

Britten, R.J., Graham, D.E., and Neufeld, B.R. (1974) Analysis of repeating DNA sequences by reassociation. *Methods Enzymol.*, **29E**, 363–420.

Chang, C.-T., Hain, T.C., Hutton, J.R., and Wetmur, J.G. (1974) The effects of microscopic viscosity on the rate of renaturation of DNA. *Biopolymers*, **13**, 1847–55.

Davis, L., Kuehl, M., and Battey, J. (1994) *Basic Methods in Molecular Biology*, 2nd edn. Paramount Publishing Business and Professional Group, Appelton & Lange, Norwalk, CT.

Feinberg, A.P. and Vogelstein, B. (1983) A technique for radiolabeling DNA restriction endonuclease fragments to high specific activity. *Anal. Biochem.*, **132**, 6–13.

Geoffrey, M.W., Berger, S.L., and Kimmel, A.R. (1987) Molecular hybridization of immobilized nucleic acids: theoretical concepts and practical consideration. In *Method in Enzymology. Guide to Molecular Cloning Techniques*, Vol. 152, S.L. Berger and A.R. Kimmel (eds), pp. 399–407. Academic Press, New York.

Höltke, H.J., Seibl, R., Burg, J., Mühlegger, J., and Kessler, C. (1990) Nonradioactive labeling and detection of nucleic acids: II. Optimization of the digoxigenin system. *Mol. Genet. Hoppe-Seyler*, **371**, 929–38.

Hutton, J.R. (1977) Renaturation kinetics and thermal stability of DNA in aqueous solutions of formamide and urea. *Nucleic Acids Res.*, **4**, 3537–55.

Kessler, C. (1992) *Nonradioactive Labeling and Detection of Biomolecules*, pp. 206–11. Springer, Berlin, Heidelberg, and New York.

Kessler, C., Höltke, H.J., Seibl, R., Burg, J., and Mühlegger, J. (1990) Nonradioactive labeling and detection of nucleic acids. I. A novel DNA labeling and detection system based on digoxigenin:anti-digoxigenin ELISA principle. *Mol. Genet. Hoppe-Seyler*, **371**, 917–27.

McCreery, T. and Helentijaris, T. (1994) Production of hybridization probes by the PCR utilizing digoxigenin-modified nucleotides. In *Protocols for Nucleic Acid Analysis by Nonradioactive Probes. Methods in Molecular Biology*, Vol. 28, P.G. Isaac (ed.), pp. 67–71. Humana Press, Totowa, NJ.

Marmur, J. and Doty, P. (1962) Determination of the base composition of deoxyribonucleic acid from its thermal melting temperature. *J. Mol. Biol.*, **5**, 109–18.

Nierzwicki-Bauer, S.A., Gebhardt, J.S., Linkkila, L., and Walsh, K. (1990) A comparison of UV crosslinking and vacuum baking for nucleic acids immobilization and retention. *BioTechniques*, **9**, 472–8.

Orosz, J.M. and Wetmur, J. (1977) DNA melting temperatures and renaturation rates in concentrated alkylammonium salt solutions. *Biopolymers*, **16**, 1183–90.

Rost, A.-K. (1995) Nonradioactive Northern blot hybridization with DIG-labeled DNA probes. In *Quantitation of mRNA by Polymerase Chain Reaction. Nonradioactive PCR Methods*, T.H. Köhler, D. Laßner, A.-K. Rost, B. Thammm, B. Pustowoit, and R. Remke (eds), pp. 93–114. Springer, Berlin and Heidelberg.

Sambrook, J., Fritsch, E.F., and Maniatis, T. (1989) *Molecular Cloning. A Laboratory Manual*, pp. 9.48. Cold Spring Harbor Laboratory Press, Cold Spring Harbor, NY.

Surzycki, S.J. (2000) *Basic Techniques in Molecular Biology*. Springer-Verlag, Berlin, Heidelberg, and New York.

Vernier, P., Mastrippolito, R., Helin, C., Bendali, M., Mallet, J., and Tricoire, H. (1996) Radioimager quantification of oligonucleotide hybridization with DNA immobilized on transfer membrane: application to the identification of related sequences. *Anal. Biochem.*, **235**, 11–19.

Wetmur, J.G. (1991) DNA probes: application of the principles of nucleic acid hybridization. *Crit. Rev. Biochem. Mol. Biol.* **26**, 227–59.

Wetmur, J.G. (1995) Nucleic acid hybrids, formation and structure of. In *Molecular Biology and Biotechnology. A Comprehensive Desk Reference*, Meyers, R.A. (ed.), pp. 605–8. VCH Publishers, Inc., New York

Wetmur, J.G. and Davidson, N. (1968) Kinetics of renaturation of DNA. *J. Mol. Biol.*, **3**, 349–70.

Yamaguchi, K., Zhang, D., and Byrn, R.A. (1994) Modified nonradioactive method for Northern blot analysis. *Anal. Biochem.*, **218**, 343–6.

FOURTH LABORATORY PERIOD

In this laboratory period you will continue the hybridization experiment. First, you will remove mismatched hybrids from the membrane using a washing procedure. Second, you will prepare the membrane for signal detection by chemiluminescence.

Protocol

Washing reaction

1. Retrieve your roller bottle from the hybridization oven and discard solution II.
2. Add 20 ml of washing solution II to the bottle and place it back into the hybridization oven. Rotate it at **maximum speed** for 10 minutes.
3. Remove the bottle from the hybridization oven and pour off and discard solution II. Drain the liquid well by placing the bottle on end on a paper towel for 1 minute.
4. Add 20 ml of washing solution III prewarmed to 65°C. Place the bottle into the hybridization oven preheated to 65°C. Rotate it at **a slow speed** for 20 minutes.
5. Pour off solution III and discard it. Drain the solution well by placing the bottle on end on a paper towel for 1 minute. Repeat the wash with solution III one more time.
6. Remove solution III and drain the roller bottle well as described above.

Preparation of membrane for detection

1. Add 20 ml of buffer A (washing buffer) to the roller bottle. Cool the oven to room temperature and rotate the bottle at **maximum speed** for 2–5 minutes.
2. Retrieve the roller bottle from the oven and discard buffer A. Add 10 ml of blocking solution (buffer B) to the bottle. Incubate for 1–2 hours **rotating slowly** at room temperature.
3. Pour off and discard buffer B. Invert the roller bottle over a paper towel and let it drain well for 1 minute.
4. Measure 10 ml of buffer B in a plastic conical centrifuge tube and add 1 µl of anti-DIG-alkaline phosphatase solution (antibody solution). Mix well and add the solution to the bottle with the membrane. Incubate for 30 minutes **rotating slowly** at room temperature. **Note:** the working antibody solution is stable for approximately 12 hours at 4°C. Do not prolong the incubation with antibody over 30 minutes. This will result in high background.

5. Pour off and discard the antibody solution. Drain the liquid well by placing the bottle on end on a paper towel for 2 minutes.
6. Add 20 ml of buffer A to the bottle and wash the membrane at room temperature rotating at slow speed for 20 minutes.
7. Add 30 ml of buffer C to the roller bottle and place it in the hybridization oven. Let it rotate at slow speed until the next laboratory period.

FIFTH LABORATORY PERIOD

In this laboratory period you will carry out the signal detection procedure and analyze data obtained in the DNA fingerprinting multi-locus experiment.

Protocol

Signal detection

1. Retrieve the roller bottle from the oven and move the membrane towards the bottle opening by gently shaking the bottle. Discard buffer C. Remove the membrane to a Pyrex dish, place on a rotary shaker, and add 200 ml of buffer C. Wash the membrane by rotating it slowly for 25 minutes. Discard buffer C and repeat the wash one more time.
2. Place a plastic bag on a sheet of Whatman 3 MM paper and add 10 ml of buffer C. Wearing gloves, transfer the membrane from the Pyrex dish into the plastic bag. Open the bag and insert the membrane into it. Place the membrane into the pool of buffer and move it with your fingers to the end of the bag. Leave as little space as possible between the membrane and the end of the bag. This will limit the amount of expensive chemiluminescent substrate necessary for filling the bag.
3. Pour off buffer C from the bag. Remove the remaining liquid gently pressing it out with a Kimwipe tissue. **Note:** do not press strongly on the membrane because this will increase the background. Most of the liquid should be removed from the bag, leaving the membrane slightly wet. At this point, a very small amount of liquid will be visible at the edge of the membrane.
4. Open the end of the bag slightly, leaving the membrane side that does not contain DNA attached to the side of the bag. Add 0.9–1 ml of CDP-Star solution directing the stream towards the side of the bag. Do not add solution directly onto the membrane.
5. Place the bag on a sheet of Whatman 3 MM paper with the DNA side up and distribute the liquid over the surface of the membrane by gently moving the liquid around with a Kimwipe tissue. Make sure that the entire membrane is evenly covered. Do not press on the membrane because this will cause "press marks" on the film. Gently remove excess CDP-Star from the bag by guiding excess solution towards the open end of the bag and onto the Whatman paper with a Kimwipe. Make sure that the membrane remains damp. **Note:** at this point small liquid droplets will be visible on the edge of the membrane, but liquid should not be present on the membrane surface. Seal the bag with a heat sealer.

6. Place the bag in an X-ray film cassette, with the DNA side of the membrane up. In a darkroom, place X-ray film over the membrane. Expose the film for 5–10 minutes at room temperature. Open the cassette and develop the film using standard procedures for film development. **Note:** maximum light emission for CDP-Star is reached in 20–30 minutes, the light emission remains constant for approximately 24 hours, and the blot can be exposed to film a number of times during this period. The best results are usually obtained when the membrane is exposed the next day.
7. After exposure, store the bag with membrane at 4°C. This membrane will be used again in experiment 3. The membrane can be stored this way for several weeks.
8. Place the acetate sheet, with the outline drawing of your gel prepared previously, on top of the X-ray film. Carefully match the top of the drawing with the outline of the membrane on the film. Mark the positions of DNA standard bands on the film with a marker pen. Analyze the date as described in the next section.

Data analysis

The most important parameter of multi-locus DNA fingerprinting is the band-sharing coefficient or similarity indices. This coefficient quantifies differences between individual DNA fragment profiles. Using these coefficients one can compare two individuals with each other. In addition, a single individual can be compared to several different populations each characterized by their unique band-sharing values. For example, average band sharing between unrelated individuals is 0.25 in both the North European and Indian subcontinent populations when *Hae*III endonuclease-digested DNA is probed with 33.15 or 33.6 probes. Thus, if the band-sharing coefficient of your DNA is 0.25 when compared to the average of North European population, you probably belong to this population.

Band-sharing coefficient D is described by the equation

$$D = 2N_{ab}/(N_a + N_b) \qquad (2.12)$$

and the similarity coefficient X is described by the equation

$$X = 1/2 \times [(N_{ab}/N_a)] + (N_{ab}/N_b) \qquad (2.13)$$

where N_{ab} is the number of scorable bands common to DNA of individuals a and b, N_a is the number of scorable bands in individual a, and N_b is the number of scorable bands in individual b.

The band-sharing values D and similarity coefficient X yield rather similar results though D is biased slightly downward as compared to the value of

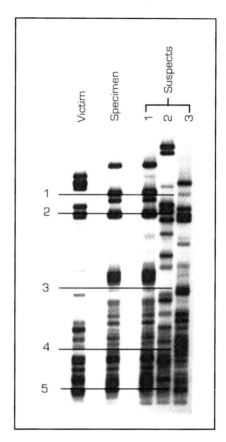

Figure 2.4 Example of multi-locus DNA fingerprinting using a 33.15 probe. DNA fingerprinting of a victim, a specimen, and three suspects is shown. Suspect 1 matches the specimen for all bands. Lines subdivide the autoradiogram into five regions for ease of analysis.

X. Knowing the D value (or X value), it is possible to estimate the chance of finding the same DNA fingerprint in two individuals. This can be calculated using the equation

$$P_n = D^n \tag{2.14}$$

where n equals the average number of scored bands per individual.

From this it is obvious that, when more bands are scored, it is less probable that they will be identical in unrelated individuals. For example, when $D = 0.25$ and nine bands are scored the probability that all of them will be identical in individuals that are not related is one in 4 million ($D^9 = 3.8 \times 10^{-6}$). For 25 bands scored the probability that they are all identical by chance is one in 10^{15} ($D^{25} = 8.8 \times 10^{-16}$).

An example of such analysis is described below. Figure 2.4 presents multi-locus DNA fingerprinting of a victim, a specimen, and three suspects. This autoradiogram shows that suspect 1 is matching the specimen for all bands. The number of bands in common that can be scored is 21. We can now calculate P_n knowing that, for this probe, the band-sharing coefficient is equal to 0.25. Thus, P_n is equal to $P_n = 0.25^{21} = 2.2 \times 10^{-13}$.

The chance that an identical pattern will be found is 4×10^{12}. Since there are only 5×10^9 human beings living, suspect 1 is the one person on earth who can have an identical pattern to the specimen. What is the probability that suspect 2 is this person? This person has five bands identical with the specimen. Thus, for suspect 2 P_n is equal to $P_n = 0.25^5$ or 9.7×10^{-4}.

This match will be found in one person per 1,024 tested ($1/9.7 \times 10^{-4}$), indicating that suspect 2 cannot be guilty of this crime.

Using the example above, calculate the chance of finding the same DNA fingerprint comparing your DNA and your partner's DNA to two standard DNAs (P_n). Assume that one of the standard DNAs given to you is the DNA of the victim and the second DNA represents a specimen found on the crime scene. The DNA fingerprints of you and your partner represent the DNA of the two suspects. Calculate the probability that you or your partner is "guilty" of this crime. The band-sharing coefficient for the M13 minisatellites family is 0.25. Can you or your partner be fully exonerated of this crime on the basis of these calculations?

CHAPTER 3

DNA Fingerprinting: Single-locus Analysis

Introduction

In this experiment you will fingerprint your DNA using a single-locus probe. To this end you will use the membrane prepared in Chapter 2. While performing single-locus analysis you will learn two procedures: removing old probe from the membrane (membrane stripping) and a hybridization signal detection method called direct detection.

First, the multi-locus probe will be stripped off the membrane. Membrane stripping procedures are frequently used in research and forensic analysis. It permits hybridization of several different probes to the same Southern blot. Next, the membrane will be hybridized with a single-locus probe and the position of the hybridized band will be visualized by chemiluminescence. The procedure you will use is a standard procedure used in forensic laboratories. In this hybridization protocol, probe is conjugated directly to reporter enzyme – alkaline phosphatase. The presence of enzyme during hybridization makes it necessary to modify standard hybridization procedures in order to protect the enzyme. This technique is frequently called the direct detection procedure and is more sensitive and faster than the indirect detection used in Chapter 2 (Surzycki, 2000).

The D2S44 single-locus probe will be used in this experiment. This probe represents a tandem repeat region present on human chromosome 2. Chromosome 2 is the result of fusion between the two ancestral primate chromosomes. The human repeat unit, with one single base substitution, is present in both the gorilla and the chimpanzee, but it is not tandem repeated. Thus, this tandem repetition arose after the evolutionary split between chimpanzees and humans (Nakamura et al., 1987; Holmlund and Lindblom, 1995; Evertsson, 1999; Holmlund, 1999).

This experiment will be carried out over two laboratory periods. In the first laboratory period, the nylon membrane will be stripped, hybridized, and exposed to film. In the second laboratory period, you will analyze the results. Figure 3.1 presents a schematic outline of the experiment.

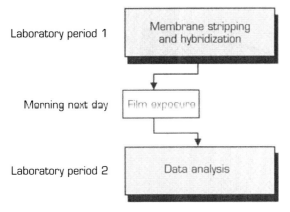

Figure 3.1 Schematic outline of the single-locus DNA fingerprinting experiment.

Background

DNA typing with mini-satellite (variable number tandem repeat or VNTR) loci is achieved by creating DNA fragment lengths containing variable numbers of repeated monomeric species. For this, genomic DNA is restricted with enzyme that cuts externally to the block of tandem repeats and analyzed using Southern blot hybridization. A Southern blot technique is essential in mini-satellite (VNTR) analysis due to a large fragment length for these loci (800 bp to several kilobases). This fragment size is outside the capability of PCR (polymerase chain reaction) techniques.

The multi-locus analysis described in Chapter 2 uses consensus core sequence as a hybridization probe. This probe, under low stringency hybridization conditions, hybridizes to a number of related VNTR sequences. Consequently, polymorphism at multiple loci is simultaneously identified and appears as a large number of bands on an autogram. The technique has its advantages and disadvantages. The advantage of this analysis is that it is highly informative because large numbers of loci are analyzed at the same time. The major disadvantages of the method are that alleles at different loci cannot be identified, thereby resulting in difficulties of statistical population analysis and it is difficult to reproduce the results due to difficulties in controlling the stringency of hybridization on which a number of the bands depend. These difficulties led to the development of a single-locus technique for analysis of VNTR allelic polymorphism.

The single-locus DNA typing (DNA fingerprinting) technique is very similar to the multi-locus method. First, DNA is digested with restriction enzyme that does not recognize the DNA sequence within the VNTR core. Next restricted DNA is hybridized by Southern blot using a specific probe that recognizes only one member of the VNTR family. The probe is usually

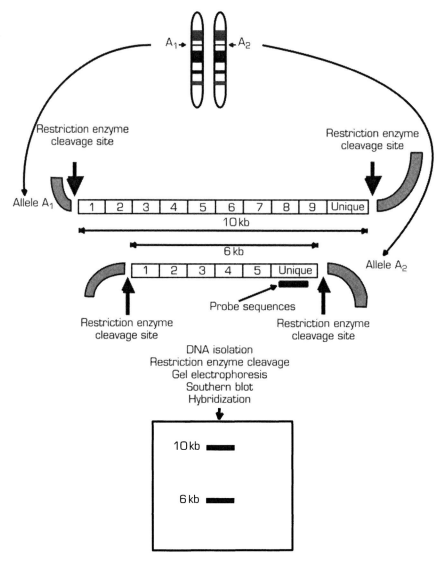

Figure 3.2 Principle of single-locus DNA fingerprinting. The single VNTR locus A is present on two homologous chromosomes at positions A_1 and A_2. The repeat of 1,000 bp is repeated nine times in allele A_1 and six times in allele A_2. Purified DNA is cut with restriction enzyme that does not have a recognition site in the repeated elements, but cuts DNA at some short distance from the beginning of the repeat (unique site). Restriction of the A_1 allele generates DNA fragments of 10 kb and restriction of the A_2 allele results in DNA fragments of 6 kb. Both A_1 and A_2 fragments contain the repeated element and DNA adjacent to it. This adjacent DNA is identical for both alleles and present only once in the entire genome. The fragments are separated by agarose gel electrophoresis and Southern blotted to a membrane. Hybridization is carried out with probe that is **not complementary** to the repeated element but to the **unique DNA** located adjacent to locus A. Hybridization generates two bands corresponding in size to two A alleles present in two homologous chromosomes.

a unique DNA sequence that is located between the end of the repeated array and the site of the restriction enzyme cut. The principle of this method is illustrated in Fig. 3.2. Consequently, as a result of hybridization, only two bands are obtained rather than the multiple bands present in multi-locus analysis. Because the single loci detected by this method are characterized, the bands represent the DNA genotype of the individual analyzed.

DNA profiling using single-locus probes is most frequently used in forensic analysis. This is because the use of these probes makes it possible to analyze data using statistical analysis of population genetics. In addition, interpretation of the results is easy since each individual has only two polymorphic variants, one inherited from the mother and one from the father. Multi-locus DNA fingerprinting analysis using single-locus probes is generated by the application of several probes (usually five to six), each one specific for a well-characterized polymorphic locus. The probes currently in use are those for the loci D1S7, D2S44, D4S139, D5S110, D10S28, and D17S79.

In the USA, the most widely used method for forensic profiling is based on a protocol developed by the FBI. It uses the enzyme *Hae*III for digesting genomic DNA samples and employs probes for single-locus VNTR markers (Budowle et al., 2000).

The locus that is used in this experiment is D2S44 VNTRs located on the long arm of chromosome 2. It has more than 30 alleles, ranging in size from 1.0 to 5.0 kb.

FIRST LABORATORY PERIOD

Protocol

Stripping the membrane

In order to hybridize the membrane with the new probe it is necessary to remove old probe from the membrane. Using this procedure one can strip and reprobe a membrane at least ten times.

1. Working in gloves, cut open the hybridization bag containing the membrane. Cut with scissors along three edges of the bag being careful not to cut into the membrane. Using forceps open the bag "like a book." Pick up the membrane with your gloved hands and roll it into a tube-like shape with the DNA side being inside.
2. Insert the rolled membrane into a roller bottle. Make sure that the side of the membrane with DNA is facing away from the glass. Do not worry that the rolled membrane is narrower than the bottle. The membrane will expand when you add liquid to the roller bottle.
3. Add 30 ml of sterile water to the roller bottle. Wash the membrane rotating at maximum speed for 5 minutes. Discard the water. This step removes CDP-Star (chemiluminescent substrate) from the membrane.
4. Add 20 ml of alkaline probe-stripping solution to the roller bottle and rotate the bottle at 37°C at maximum rotation speed for 10 minutes. Discard the alkaline probe-stripping solution. Invert the bottle over a paper towel and let the remaining liquid drain for 1 minute.
5. Add 20 ml of two times SSC solution to the roller bottle and wash the membrane at room temperature rotating the bottle at maximum speed for 20 minutes. Discard the solution and repeat the washing procedure one more time.

Pre-hybridization

1. Add 10 ml of hybridization solution to a 15 ml plastic centrifuge tube and preheat it for 10 minutes in a 55°C water bath. Add the solution to the roller bottle.
2. Close the roller bottle tightly and place it in the hybridization oven. Rotate slowly at 55°C for 20 minutes.
3. Start preparing the probe for hybridization. Pour 10 ml of hybridization solution into a 15 ml conical centrifuge tube. Add 1 µl of probe to the tube and incubate it at 55°C in a water bath until it is ready to add to the roller bottle. **Note**: your instructor will give you the D2S44-labeled probe.
4. After pre-hybridization has been completed, remove the bottle from the oven and pour off the pre-hybridization solution.

5. Add prewarmed hybridization solution prepared in step 3 to the roller bottle. **Note**: it is very important that prewarmed hybridization solution is added to the bottle as fast as possible and the bottle returned to the oven immediately.
6. Return the bottle to a hybridization oven and allow it to **rotate slowly** at 55°C for 20 minutes. Remember that the hybridization temperature is critical for specific hybridization and it must be 55°C for at least 20 minutes.
7. Take your bottle from the hybridization oven and pour off the hybridization solution into a 15 ml centrifuge tube. Return the tube with the probe to your instructor. This probe can be reused several times.
8. Drain liquid from the bottle by inverting it over a paper towel for 1–2 minutes.

Washing reaction

1. Add 20 ml of wash buffer 1 to the tube. Place the tube into the hybridization oven and allow it to rotate slowly for 10 minutes at 55°C. Repeat the washing procedure one more time.
2. Remove the roller bottle from the oven and pour off wash buffer 1. Drain the solution well by inverting the bottle over a paper towel for 1–2 minutes.

Signal detection

1. Add 20 ml of buffer C (detection buffer) to the roller bottle. Close the bottle and move the membrane towards the tube opening by gently shaking the tube.
2. Open the bottle and remove the membrane to a Pyrex dish filled with 200 ml of buffer C (detection buffer). Wash the membrane for 20 minutes on a rotary shaker.
3. Place a plastic bag on a sheet of Whatman 3MM paper and add 10 ml of buffer C. Wearing gloves transfer the membrane from the Pyrex dish into the plastic bag. Open the bag and insert the membrane into it. Place the membrane into a pool of buffer and move it with your fingers to the end of the bag. Leave as little space as possible between the membrane and the end of the bag. This will limit the amount of expensive chemiluminescent substrate needed.
4. Pour off buffer from the bag. Remove the remaining liquid by gently pressing it out with a Kimwipe tissue. **Note**: do not press strongly on the membrane because this will increase the background. Most of the liquid should be removed from the bag leaving the membrane slightly damp. At this point a very small amount of liquid will be visible at the edge of the membrane.
5. Open the end of the bag slightly, leaving the membrane side that does not contain DNA attached to the side of the bag. Add 0.9–1 ml of CDP-Star

solution directing the stream towards the side of the bag. Do not add solution directly onto the membrane.

6. Place the bag on a sheet of Whatman 3MM paper with the DNA side up and distribute the liquid over the surface of the membrane by gently moving the liquid around with a Kimwipe tissue. Make sure that the entire membrane is evenly covered. Do not press on the membrane because this will cause "press marks" on the film. Gently remove excess CDP-Star from the bag by guiding excess solution towards the open end of the bag and onto the Whatman paper with a Kimwipe tissue. Make sure that the membrane remains damp. **Note**: at this point, small liquid droplets will be visible on the edge of the membrane, but liquid should not be present on the membrane surface. Seal the bag with a heat sealer.

7. Place the bag in an X-ray film cassette DNA side up. In a darkroom, place X-ray film over the membrane. Expose the film for 5–10 minutes at room temperature. Open the cassette and develop the film using standard procedures for film development. **Note**: maximum light emission for CDP-Star is reached in 20–30 minutes, the light emission remains constant for approximately 24 hours, and the blot can be exposed to the film a number of times during this period. The best results are usually obtained when the membrane is exposed the next day.

SECOND LABORATORY PERIOD

Protocol

To analyze your data, follow the protocol and do the calculations as shown in the example shown in the section on expected results.

1. Align the X-ray film with the gel drawing made previously. The outline of the filter should be visible on the film. Measure the electrophoretic mobility of each band.
2. Prepare a semi-log graph of the standard DNA fragments. Use the plot of electrophoretic mobility of the known size fragments as a function of the log of fragment size in base pairs. The instructor will give you the sizes of the standard DNA fragments.
3. Measure the electrophoretic mobility of the DNA fragments that hybridize to the probe. Determine the sizes of these fragments using the graph prepared in step 2.
4. Using DNA band size, find to which bin(s) your allele belongs. Use Table 3.1 for this determination. Read the frequency of the bin to which each of your bands belongs from Table 3.1.
5. Prepare the uncertainty windows table for all bands. Next prepare a match windows table for your bands and calculate the probability that a randomly chosen person from the population will have the same profile as you. For all these calculations use the example described in the section on expected results.

Data analysis

This data analysis will simulate the single-locus analysis used in forensic cases. The DNA of control 1 will be taken as victim DNA. The DNA of the second control will be treated as DNA found at the crime scene. Both these DNAs are DNA of **evidentiary samples**. Your DNA and the DNA of your partner will be the DNA of two suspects.

Your task will be to calculate the statistical probability that the DNA profile of any one person will contain all the alleles represented by the bands of the evidentiary samples. Thus, the question addressed by this DNA analysis is what is the probability that a person chosen at random from the relevant population will likewise have a DNA profile matching that of the evidentiary sample? You will therefore determine what is the probability that your DNA single-locus profile is matching the "evidentiary sample," i.e. are you "guilty" or not of this imaginary crime.

The first step of this analysis is to compare the DNA patterns on your autogram in order to determine whether your DNA pattern matches the DNA pattern of the evidentiary samples. The patterns are visually

Table 3.1 Allelic frequencies of the D2S44 VNTR locus for DNA cut with *Hae*III enzyme

Bin number	Size range (bp)	N	Frequency
3	0–871	8	0.005
4	872–963	5	0.003
5	964–1,077	24	0.015
6	1,078–1,196	38	0.024
7	1,197–1,352	73	0.046
8	1,353–1,507	55	0.035
9	1,508–1,637	197	0.124
10	1,638–1,788	170	0.107
11	1,789–1,924	131	0.083
12	1,925–2,088	79	0.050
13	2,089–2,351	131	0.013
14	2,352–2,522	60	0.038
15	2,523–2,692	65	0.041
16	2,693–2,862	63	0.040
17	2,863–3,033	136	0.086
18	3,034–3,329	141	0.089
19	3,330–3,674	119	0.075
20	3,675–3,979	36	0.023
21	3,980–4,322	27	0.017
22	4,324–5,685	13	0.008
25	5,686–	13	0.008

evaluated in order to determine a likely match. At this step a suspect may be excluded from further consideration because the pattern will be noticeably different from the evidentiary samples. If a suspect pattern is close to the evidentiary samples (even for one band), this is considered a "likely match."

If a suspect's bands cannot be excluded, they are subjected to analysis in order to determine the length of the represented DNA fragments measured in base pairs. These measurements are taken by comparing the bands for the sample fragments with the bands for the size marker fragments of known base pair lengths.

The most straightforward means of making a "likely match" calculation is to apply the product rule to **allele frequencies** derived from major population groups. The essence of the product rule is multiplication of individual band probabilities in order to arrive at an overall probability, statistically expressed as a simple fraction, that one out of a stated number of persons in the population (e.g. one out of 100,000) will match the DNA profile of the evidentiary sample in question.

In court procedures, at least six to seven independent single-locus alleles are analyzed. Thus, the rule is applied in two stages: first, to determine the allelic frequency at each locus and then to determine the alleles' combined frequency at all loci.

Measurement of band sizes using gel electrophoresis is not exact because the resolving power of the gel is limited. This results in an inability to separate alleles that are close in size. For comparison purposes therefore the database bands are sorted into ranges of size called "bins." There are two kinds of bins: "floating bins" and "fixed bins." Forensic analysis in the USA uses a fixed bins database of allele frequencies prepared by the FBI.

The fixed bins database contains a catalog of the entire spectrum of VNTR base pair sizes likely to appear as bands. A separate fixed bin table has been compiled for each locus. Each band is recorded within the bin that encompasses its base pair size. To protect a suspect against unduly small **frequencies**, any bin with four or fewer bands is combined with its neighbor until each bin contains a minimum of five bands.

Table 3.1 presents the bin size and allelic frequencies of the D2S44 locus for *Hae*III-restricted human genomic DNA. The fixed bin table shows not only each bin's range of sizes and number of bands (N), but also each bin's frequency, which is calculated from the ratio of the number of bands in the bin to the total number of bands in the table. The FBI laboratory prepared this table.

In fixed bin analysis the frequency of an evidentiary band is determined by assigning it the frequency of the fixed bin into which its base pair size falls. In order to calculate the frequency (probability) of occurrence of a particular allele arrangement for a heterozygote (two bands), the **frequencies** of those bands are multiplied by each other and the result multiplied by 2.0 ($2pq$). This is because the frequency of heterozygotes in a Hardy–Weinberg (random) population is equal to $2pq$, where p is the frequency of one allele in question and q is the frequency of the second allele. If there is only one band at the locus, either the donor is homozygous or there is a second **allele** that for some reason did not appear. In order to take both possibilities into account while avoiding prejudice to the suspect, the frequency of the first band is simply multiplied by 2.0.

The next step is to calculate the probability that a randomly chosen person has the same profile as the evidence sample. For this the "product rule" is applied. The frequencies of matched large and small bands for each of the probes are multiplied by each other and multiplied by 2.0 in order to determine the frequency of heterozygotes ($2pq$). This value is calculated for each probe and multiplied by each other. For example, let us assume that the matched frequency for probe 1 is 0.017, for probe 2 is 0.26 and 0.015, for probe 3 is 0.072 and 0.013, for probe 4 is 0.089, for probe 5 is 0.017, and for probe 6 is 0.089. Thus, the overall probability is equal to $2(0.017) \times 2(0.26)(0.015) \times 2(0.072)(0.013) \times 2(0.089) \times 2(0.017) \times 2(0.089) = 1.6 \times 10^{-11}$ or one in 60 billon. Thus, the probability of a random match with the evidentiary samples other than the suspect is very small because there are only 5 billion people living on earth.

Expected results

Figure 3.3 presents the autogram of single-locus analysis similar to your autogram. S1 and S2 are two suspects (your DNA and your partner's DNA) and E1 and E2 are evidentiary samples (control DNA 1 and 2).

The results presented in Fig. 3.3 indicate that both "suspects" cannot be excluded and represent a "match" because the bands in S1 and S2 are located very close to at least one of the bands in the evidentiary samples. The next step of analysis is to determine the sizes of all of the bands for the S1, S2, E1, and E2 data. The results are recorded in the size column of the uncertainty windows table (Table 3.2).

Next, the upper and lower limits of the uncertainty windows for each band are calculated. The size of these windows spans the range 2.5 percent below to 2.5 percent above the measured value. To calculate this value, the band size is multiplied by 0.025 and recorded in the column named 2.5 percent in Table 3.2. This value is added and subtracted to the corresponding band size and both values are recorded in the column named uncertainty window.

Subsequently, a comparison is made between the uncertainty windows for each suspect band and those of the evidentiary bands. For example, comparing the uncertainty window for a large band of S1 with windows for E1 shows that only one window of the E1 band (band 2) overlaps. Comparing the uncertainty window for the small S1 band with E1 windows shows that none of the windows overlap. Furthermore, a comparison of both S1 bands with the band of E2 also indicates no overlap.

This analysis is continued for suspect 2 (S2). In this example there are two overlaps for this subject. The uncertainty widow for the small band of S2 overlaps windows for band 4 of E2 and the small band of E2 (see Table 3.2).

The next step is to calculate the size of the match window that will be used for finding the frequency of these markers in the database for the allele frequency of D2S44 presented in Table 3.1. For this analysis, match windows are calculated with 5 percent error. These calculations are performed for each band of the suspect that overlaps bands of the evidentiary samples. The results are recorded in a table. In the above example, only two bands are analyzed. These are overlap of the large band of S1 with band 2 of E1 (4,549) and the overlap of the small band of S2 with two bands of S2 (small band and band 4). Table 3.3 shows the results of this calculation for the example given.

In order to calculate the "match window" the size of the large band of S1 (4,549) is multiplied by 0.05 and recorded in the 5 percent column of Table 3.3. The "match window" is calculated by the addition and subtraction of the 5 percent value to and from the size of the large band. The results are recorded in the match window column. Match window values are used for finding the bin number in Table 3.1 that overlaps the match window value.

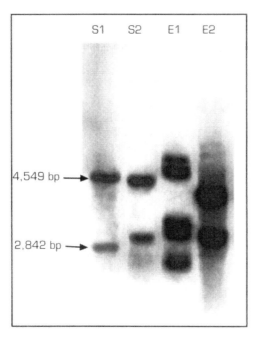

Figure 3.3 Autogram of single-locus analysis. Autogram of the DNA of two suspects and two evidentiary samples. S1, suspect 1; S2, suspect 2; E1, evidentiary DNA 1; E2, evidentiary DNA 2.

In the example, two bins overlap the value of the match window, i.e. bins 21 and 22. This is recorded in the bin(s) column. The frequencies for these bins are 0.017 and 0.008, respectively (see Table 3.1). Only the larger of these two frequencies is recorded in the table. The results of this analysis for the S2 suspect are also shown in Table 3.3.

In forensic analysis, the probe will be "stripped" from the membrane and then hybridized with the next single-locus probe. The results are calculated for the match with this probe. This process is repeated with at least six other probes.

In our example, we analyzed the result of hybridization with only one probe. Thus, the probability of a match (or non-match) is high. For S1 individuals, only one band matched the evidentiary sample. To calculate this probability, we multiply the frequency for a single band by 2.0. Thus, the probability is 2 × 0.017 = 0.034. This means that if we analyze DNA from 1,000 persons from any population, 34 of them will show a match with this single evidentiary band. This certainly does not exclude S1 as the suspect in this crime. However, analysis of six more single-locus probes, as required by courts, might prove S1 innocent.

For the S2 individual, this probability is 2 × 0.089 = 0.178 meaning that out of 1,000 persons from any population, 178 of them will show a match.

Table 3.2 The uncertainty windows data

Source	Band	Size (bp)	2.5% error	Uncertainty window
S1	Large	4,549	114	4,435–4,663
	Small	2,840	71	2,911–2,769
S2	Large	4,250	106	4,356–4,144
	Small	3,000	75	3,075–2,925
E1	Band 1	5,258	131	5,389–5,127
	Band 2	4,690	117	4,807–4,573
	Band 3	3,246	81	3,327–3,165
	Band 4	3,040	76	3,116–2,964
	Band 5	2,461	61	2,522–2,400
E2	Large	4,000	100	4,100–3,900
	Small	3,082	77	3,159–3,005

Table 3.3 Match windows and frequencies of D2S44 alleles

Source	Band	Size (bp)	5%	Match Window	Bin(s)	Frequency
S1	Large	4,549	227	4,322–4,776	21 and 22	0.017
S2	Small	3,000	150	3,150–2,850	16, 17, 18	0.089

Excerpt from the court procedures of the Simpson Trial D2S44 Probe

Mr Harmon: Okay. Mr Sims, yesterday you had described A16, which is the autorad for the probe D1S7 and today we have A17 which is the autorad for **D2S44** for the membrane AM626 and could you just describe again the samples that are up there and we will go through them and I will have you mark them as best you can.

Mr Sims: Okay. Just to reorient you again, in lane 1 this is one of the ladders, the size standards that we mentioned.

Mr Harmon: Could you get a different color for that. That will not show up really well.

Mr Sims: Is that better?

Mr Harmon: Sure.

Mr Sims: Again, this is lane 1. This is the size standard. Lane 2 is the K562. That is the national standard that we use. Lane 3 is the quality control, the blind sample. Lane 4 is another size standard. Lane 5 and the next two lanes now I will talk about, these are the reference bloodstains from Nicole Brown and then next we have the reference bloodstain from Ronald Goldman.

Mr Harmon: Okay. And is it easy to distinguish between Miss Brown and Mr Goldman at their – by their reference sample?

Mr Sims: Yes, it is.
Mr Harmon: You do not need a computer to do that?
Mr Sims: No, you do not.

Day 3

Mr Harmon: Could you – just the way we approached the other statistics, could you describe from among the three major population groups that you have used in your calculations the frequency estimate for those three groups and the match between Nicole Brown's reference blood and Greg Matheson's cut-out stain on the sock?
Mr Sims: Yes. For those – the six loci that I looked at, these are RFLP loci D1S7, **D2S44**, D4S139, D5S110, D10S28, and D17S79, the profile detected in stain 42-A(1) occurs in approximately 1 in 21 billion Caucasians, 1 in 41 billion African Americans and 1 in 7.7 billion Hispanics, again indicating that this profile is a rare event and pointing out that these are for unrelated individuals.

References

Budowle, B., Smith, J., Moretti, T., and DiZinno, J. (2000) *DNA Typing Protocols: Molecular Biology and Forensic Analysis*. BioTechniques Book Publication, Eaton Publishing, Natick, MA.

Evertsson, U. (1999) The minisatellite D2S44 in chimpanzee has a different tandem repeated unit localized 200 bases upstream of the human repeat array. *Student J. Health Sci.*, **2**, 13.

Holmlund, G. (1999) The polymorphism of the minisatellite system D2S44. Medical dissertation, Linkoping University.

Holmlund, G. and Lindblom, B. (1995) Flanking region sequences and internal repeat structure of the pYNH24 (D2S44) 2 kbp insert analyzed by polymerase chain reaction and partial digestion with *Rsa*I. *Electrophoresis*, **16**, 1881–5.

Nakamura, Y., Gillian, S., O'Connell, P., Leppert, M., Lathrop, G., Lalouel, J.-M. et al. (1987) Isolation and mapping of a polymorphic DNA sequence pYNH24 on chromosome 2 (D2S44). *Nucleic Acids Res.*, **15**, 10073.

Surzycki, S. (2000) *Basic Techniques in Molecular Biology*. Springer-Verlag, Berlin, Heidelberg, and New York.

General reading

(1998) *The Evaluation of Forensic DNA Evidence. Committee of DNA Forensic Science: An Update*. National Academic Press, Washington, DC.

CHAPTER 4

Out of Africa: Origin of Modern Humans

Introduction

The goal of this experiment is to analyze linkage disequilibrium between the *Alu* locus in the CD4 gene on chromosome 12 and the short tandem repeat (STR) locus located near by. We will calculate the extent of linkage disequilibrium between these genes for the whole class "population" and compare it with known linkage disequilibrium calculated for different non-African populations. Each student will determine his or her haplotypes using a PCR (polymerase chain reaction) and determine to which world populations they most probably belong.

During the course of this experiment, students will learn how to perform a PCR, use the thermal cycler, and analyze products using high-resolution agarose gel electrophoresis.

This experiment will take two laboratory periods. During the first laboratory period students will assemble and run a PCR. The second laboratory period will be dedicated to agarose gel electrophoresis and analysis of the results. A schematic outline of the experiment is shown in Fig. 4.1.

Background

Origin of humans

Humans arose from an ancestor common to contemporary primates that were present on earth for at least 35 million years. We diverged from the ancestor of our closest primate relatives, the chimpanzees, approximately 4–5 million years ago. These humanoids, which still retained an ape-like body shape, spread across all of Africa and probably the entire world. They are represented by ardipithecines and the more specialized australopitheciones.

The first evidence of animals that are more human-like also came from Africa and they date from approximately 2 million years ago (*Homo ergaster*).

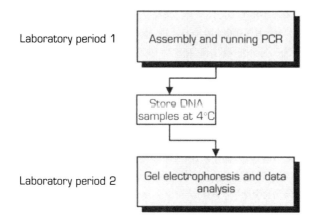

Figure 4.1 Schematic outline of experiment 4.

These animals or probably a variant of them, *Homo erectus*, emigrated from Africa approximately 1.7 million years ago as evidenced by finding *H. erectus* skeletons all over Eurasia (e.g. China and Java).

These findings led to a model of modern human evolution named the **multiple-origin model**. This theory states that major human races split one from another at the time of *H. erectus* dispersal from Africa. These groups evolved more or less independently (with some gene mixing) to modern humans (*Homo sapiens*). This version of our origin appears not to be true and has no serious advocates today.

The second hypothesis is known as the **out-of-Africa replacement (OAR)** hypothesis and is widely accepted today. In this model of human evolution, anatomically modern humans (*H. sapiens*) evolved first in Africa, probably from *H. erectus*, approximately 150,000–200,000 years ago. A small part of this group colonized Eurasia and the rest of the earth by migrating from Africa approximately 100,000 years ago driving local populations of *H. erectus*-like species to complete genetic extinction (**replacement**) without any or very little gene exchange. The migration from Africa may have occurred in a single wave or in multiple waves.

Evidence for the OAR hypothesis comes from three lines of reasoning. First is the observation of a large genetic diversity in African populations versus the rest of the world. This indicates that the African population is the oldest and has had more time to accumulate genetic variations. Large genetic variability in African populations was shown to exist for mitochondrial DNA, Y chromosome micro-satellites, and autosomal mini-satellites.

Second, phylogenetic analysis of nearly every genetically variant locus shows evolutionary trees the first branch of which separates African from non-African. This indicates that every polymorphic variant studied can be traced back to the African population. For example, well-known studies of maternally inherited mitochondrial DNA indicate that African sub-Sahara populations have much higher genetic diversity than non-Africans. Moreover, each type of mitochondria of non-African populations can be traced back through the maternal lineage to one of African diversity, i.e. to ancestral females (Eve?) that existed in Africa between 100,000 and 300,000 years ago (Cann et al., 1987; Richards et al., 1996). The interpretation of the original "mitochondrial Eve" data has been hotly debated, but recent reinterpretation of these data and the emergence of some new data based on DNA sequencing confirms the original interpretation. Similar results were obtained for Y chromosome polymorphic sites. Since this chromosome is inherited thought the paternal line, the analysis also traces the origin of males (Y chromosome – Adam?) to the African population of humans (Semino et al., 1996; Casalotti, et al., 1999; Seielstad et al., 1999).

The third evidence comes from analysis of the age of the "genetic" ancestors of the non-African human population. This population is very young, being separated from the African population approximately 100,000 years ago (emigration from Africa).

In our experiment we will use one of the most ingenious experimental proofs existing for the OAR hypothesis introduced by Tishkoff et al. (1996). These authors used decay of linkage disequilibrium between two closely linked genes on chromosome 12 to prove the out-of-Africa hypothesis.

Linkage disequilibrium is defined as the non-random association between two polymorphic loci (e.g. A and B) on the same chromosome. The term "polymorphic" (meaning many forms) describes the state of a single genetic locus that exists in two or more forms at significant frequencies in the population (above 1 percent). When mutation of one of these loci occurs, e.g. of gene A, this mutation creates another polymorphic form of this gene. This mutation occurs on a particular chromosome on which a specific polymorphic variant of another gene, gene B, resides. Thus, these two specific polymorphic variants of both genes are now together on the same chromosome and are transferred from one generation to another more frequently together (no recombination) than separately (recombination). They are said to be in linkage disequilibrium. This association will decay with successive generations as recombination breaks the original association.

The rate of the breakage of this association (linkage disequilibrium disassociation) depends on the distance between both loci and the number of successive generations. Thus, it is possible to reconstruct recent evolutionary history using such closely linked markers.

In our experiment, you will use two linked markers located 9.8 kb apart. One of these markers is the *Alu* short interspersed nuclear element (SINE) and the other is an STR. *Alu* elements are approximately 300 bp in length and derive their name from a single recognition site for the endonuclease *Alu*I, which is located near the middle of it. The *Alu* element is the most abundant sequence in the human genome. The human genome contains approximately 1 million *Alu* repeats comprising an estimated 10 percent of the genome. They are predominantly located in non-coding, intragenic locations, particularly in introns. The *Alu* element is thought to be derived from the 7 SL RNA gene that encodes the RNA component of a signal recognition particle that functions in protein synthesis.

The *Alu* element polymorphism used in our study resulted from the deletion of the *Alu* in locus CD4 located on chromosome 12. This element is present at this position in all primates and in approximately 82 percent of the chromosomes (haplotypes) in the African population. This indicates that the *Alu*(+) haplotype is the ancestral state and deletion of *Alu* (*Alu* polymorphism) occurred after divergence of humans from the great apes.

The second polymorphic marker that we will be using is an STR polymorphism consisting of a tandem repeated block of five nucleotides TTTTC. This locus has 12 alleles that differ by the number of repeats (four to 15). Only three of these alleles are seen in the human population outside of Africa at a frequency greater than 10 percent. These are STR 85 (five repeats), STR 90 (six repeats), and STR 110 (ten repeats) alleles. All of the great ape species have very limited polymorphism at this locus (three or six repeats), indicating that most of the STR polymorphic human alleles appeared after humans diverged from the great apes.

In this experiment oligonucleotide primers flanking the *Alu* insertion site will be used for amplifying a 661 bp fragment when *Alu* is present and a 438 bp fragment when it is deleted. Because humans are a diploid organism, containing two chromosome 12s, one from the father and the other from the mother, three genotypes are possible.

1. Homozygotic for the absence of *Alu*. In this case both chromosome 12s do not have the *Alu* locus. There will be only one PCR product, which is 438 bp long.
2. Homozygotic for the presence of *Alu*. In this case both chromosome 12s contain the insertion. There will be only one PCR product, which is 661 bp long.
3. Heterozygotic for the presence of *Alu*. There will be two PCR products, one which is 438 bp long and the other which is 661 bp long.

All three genotypes can be distinguished from each other following agarose electrophoresis of the PCR products.

The STR marker will be amplified using two primers (TT-F and TT-R) that are approximately 30 bp from the STR locus. Figure 4.2 shows the relative positions of the *Alu* and STR on chromosome 12.

Initially in our evolution, the *Alu*(−) allele was generated by the deletion of *Alu* on the chromosome that contained the "ancestral" STR 90 (six repeats). Thus, at this time there was total linkage disequilibrium between these alleles. In time, due to mutations and recombination, this association was lost in the ancestral African population. However, when some of the ancestral African population left Africa and populated the rest of the world most of them carried chromosome 12 containing the *Alu*(−) STR 90 haplotype. Thus, this combination of alleles (haplotype) shows almost

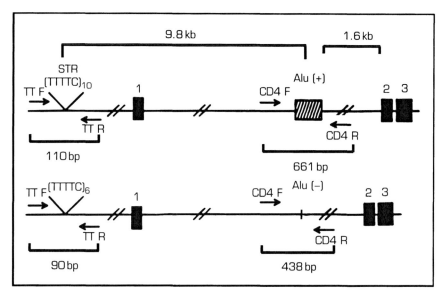

Figure 4.2 Positions of polymorphic markers at the CD4 locus on chromosome 12. Two haplotypes (chromosomes) are shown: *Alu*(+) STR with ten repeats (110 bp) and *Alu*(–) STR with six repeats (90 bp). Arrows indicate the positions of the *Alu* primers and STR primers. Solid blocks represent exons of the CD4 gene.

complete linkage disequilibrium in the non-African populations. Two other haplotypes, *Alu*(+) STR 85 and *Alu*(+) STR 110, were also present in a small population that migrated from Africa and are now distributed throughout the world.

The non-random association between the CD4 STR alleles and *Alu*(–) allele (linkage disequilibrium) can be calculated by estimating the excess of ancestral haplotype *Alu*(–) and any one of the STR alleles in non-African populations. This value is usually called P_{excess} or δ and is equal to

$$\delta_{-,x} = \frac{P_{-,x} - P_{+,x}}{1 - P_{+,x}} \tag{4.1}$$

where $P_{-,x}$ is the frequency of haplotype *Alu*(–) and any STR allele (e.g. STR 110) and $P_{+,x}$ is the frequency of haplotype *Alu*(+) and the same STR allele as above (e.g. STR 110). The δ values for world populations are presented in Table 4.1 and Table 4.2 presents the frequency estimates for Alu STR haplotypes for different world populations.

PCR

The PCR is a powerful method of *in vitro* DNA synthesis. Large amounts of a specific segment of DNA of defined length and sequence can be

Table 4.1 Values of δ for the three *Alu*(–) STRs in some world populations

Population	STR 85	STR 90	STR 115
Non-African			
Combined	–	0.98	–
Northeast Africa			
Egyptian	–	1.0	–
Ethiopian Jewish	–	0.61	–
Somalian	0.18	0.67	–
Sub-Sahara Africa			
Woloff	0.41	0.22	–
Mbuti	0.46	0.19	0.18
Bantu speakers	–	0.26	0.29
Kikuyu	–	0.11	0.64
Herero	0.01	0.38	0.13
Biaka	0.38	–	–
Nama	–	0.27	0.18

synthesized from a small amount of template. The technique has revolutionized molecular biology and is used in virtually every area of natural sciences and medicine.

The principal of a PCR is rather simple and involves enzymatic amplification of a DNA fragment flanked by two oligonucleotides (primers) hybridized to opposite strands of the template, the 3′-ends of which are facing each other. DNA polymerase synthesizes new DNA starting from the 3′-end of each primer. Repeated cycles of heat denaturation of the template, annealing of the primers, and extension of the annealed primers by DNA polymerase results in amplification of the DNA fragment. The extension product of each primer can serve as a template for the other primer, resulting in essentially doubling the amount of the DNA fragment in each cycle. The result is an exponential increase in the amount of a specific DNA fragment defined by the 5′-ends of the primers.

The products of a PCR will include, in addition to the starting DNA, an amount of a specific target sequence that can easily be visualized as a discrete band of a specific size by agarose gel electrophoresis. The practical consequence of the chain reaction is that one can start with nanogram amounts of DNA, carry out a PCR, and then run the sample on a gel in order to visualize a specific band corresponding precisely to the distance (including the length of the primers themselves) between the two primers used.

PCR kinetics

During the PCR reaction, products from one cycle serve as a template for DNA synthesis in the next cycle. Theoretically, the amount of product

Table 4.2 The percentage of *Alu* STR haplotypes for some world populations

	All STR		STR 80		STR 85		STR 90		STR 100		STR 110		STR 115		STR 120		STR 125		STR 130	
	Alu+	Alu−	Alu+	Alu−	Alu+	Alu−	Alu+	Alu−	Alu+	Alu−	Alu+	Alu−	Alu+	Alu−	Alu+	Alu−	Alu+	Alu−	Alu+	Alu−
New World	98	2.0	–	–	55.3	–	0.1	1.8	–	–	42.1	–	0.4	–	0.1	–	–	–	–	–
Pacific Island and Australo-Melanesian	98.7	1.3	–	–	72.7	–	1.3	0.7	–	–	22.7	–	1.3	–	–	–	–	–	–	–
Asian	95.1	4.9	–	–	60.3	0.1	1.1	3.7	0.3	–	31.0	–	2.3	–	0.2	–	–	–	–	–
European	72.4	27.6	–	–	36.1	0.1	1.0	27.6	–	–	28.9	0.1	3.7	–	–	–	–	–	–	–
Middle Eastern	68.0	32.0	–	–	33.5	0.5	1.5	30.5	–	0.5	24.0	–	4.5	–	2.0	–	2.5	–	–	–
Northeast African	79.2	20.8	–	–	27.3	2.0	1.0	16.0	–	0.4	28.0	–	11.3	–	6.3	–	0.6	–	2.7	–
Sub-Sahara African	82.2	17.8	0.2	–	18.7	4.6	2.3	4.2	7.4	0.7	17.2	–	15.6	5.4	10.7	0.7	2.4	0.5	2.1	0.1

doubles each cycle making the PCR process a true chain reaction, which is described by the equation

$$N = N_0 \times 2^n \qquad (4.2)$$

where N is the number of amplified molecules, N_0 is the initial number of molecules, and n is the number of amplification cycles. This equation holds true if the efficiency of amplification (E), which is defined as the fraction of template molecules that take part in amplification during each cycle, is 1.0. An equation that describes the amplification process better, taking into account the efficiency of the process, is

$$N = N_0 \times (1 + E)^n \qquad (4.3)$$

where E is the amplification efficiency.

Experimentally the accumulation of product during the course of the reaction is far from the case described by equation (4.3). This is primarily the result of changes in the amplification efficiency during the course of the reaction. At the beginning, the efficiency of amplification is close to 1.0 (0.8–0.97) and accumulation of the product proceeds exponentially. This constitutes the exponential phase of the reaction. During late PCR cycles accumulation of the product slows down and eventually stops. This effect is usually referred to as the "plateau effect" and occurs after accumulation of 0.3–1.5 pmol of amplification product (Innis et al., 1990; Sardelli, 1993).

The number of cycles it takes to reach a plateau depends mainly on the initial number of molecules present in the reaction (N_0). However, a plateau can be influenced by a number of conditions, such as the following.
1. Inhibition of polymerase by pyrophosphate.
2. Utilization of substrate.
3. The temperature of cycling.
4. The stability of enzyme and reagents.
5. The presence of impurities.
6. Reannealing of product at high concentration.
7. Incomplete strand separation at high concentrations of product.
8. Formation of primer dimers.
9. The GC content of the template.

Reaching a plateau should be avoided since it increases the probability of obtaining non-specific amplification products. Choosing the correct number of PCR cycles is one way of avoiding false initiation products. This can be achieved by correctly choosing the initial concentration of target DNA molecules. Since most standard amplification protocols use 30 amplification cycles, varying the concentration of target DNA is most frequently employed as a means of avoiding reaching a plateau.

It is important to emphasize that failure to obtain amplification products commonly occurs due to too high a concentration of DNA in the reaction. This is because the plateau effect is for the most part a consequence of a high DNA concentration in the reaction. This can come about from *de novo* DNA synthesis or from too high a concentration of template (N_0 in equation (4.3)).

Rapid cycle DNA amplification

DNA amplification requires temperature cycling of the sample. This is usually performed using a thermal cycler instrument. Reactions of denaturation of DNA and annealing of primers are very fast kinetic reactions requiring only a few milliseconds to occur at the DNA and primer concentrations used in a PCR. Most of the commercial instruments utilize metal blocks for thermal equilibration and samples are contained in plastic microfuge tubes. In these instruments very fast temperature changes are not possible. A significant fraction of the cycle time in these machines is spent on heating and cooling of the blocks and tubes and the liquid contained in them. This extends amplification times to several hours, a much longer time than is kinetically necessary for carrying out all the steps of an amplification reaction.

Moreover, extended amplification times and long transition times make it difficult to determine the optimal temperature and times for each stage of the PCR reaction. This results in many false initiations by polymerase and, consequently, leads to poor product specificity. Rapid temperature transitions and small sample volumes improve both specificity and the time necessary for carrying out PCR reactions.

The instrument that makes it possible to carry out a PCR at speeds close to those theoretically required is an air thermal cycler (Wittwer and Garling, 1991; Wittwer et al., 1994). This rapid thermal cycler instrument uses heat transfer by hot air to samples contained in thin, glass capillary tubes. Because of the low heat capacity of air and the thin walls and increased surface area of capillary tubes, samples can be heated and cooled very quickly. The total amplification time in these instruments for 30 PCR cycles is 15–20 minutes and the volume of reaction is 10 µl. This makes it an ideal instrument for teaching applications. The specificity of amplification is also dramatically increased as compared to standard instruments and the cost of the instrument is low compared to standard thermal cyclers. We will be using this instrument for our amplification reactions.

FIRST LABORATORY PERIOD

In this laboratory period you will assemble and run two PCR reactions. The first reaction will amplify the region of chromosome 12 where the *Alu* SINE element is located. The second reaction will amplify the region of chromosome 12 where the polymorphic STR locus resides (see Fig. 4.2 for details).

Technical tips

The failure to amplify product or very weak amplification usually results from a failure of some component of the reaction. Most frequently this is due to primer degradation. Special care should be taken to prevent this degradation.

Using too high a concentration of DNA template is the second most frequent cause of reaction failure. To amplify human DNA sequences, the concentration of genomic DNA in the reaction should not exceed 100 ng. Careful determination of DNA concentration in the sample is paramount to successful amplification of DNA sequences.

The ability of a PCR for amplifying minute amounts of DNA template requires unusual care in order to avoid contamination with none-target DNA. Each pipette tip should be used only once and use of tips with barriers in order to avoid contaminating the pipettors, reagent stocks, and reaction is strongly recommended. In addition, to further avoid contaminating stock solutions, template DNA should be always added to the reaction mixture last.

In order to avoid DNase contamination of the reagents gloves should be worn all of the time and all buffer solutions should be autoclaved. Moreover, each experiment should always include a negative control reaction without template DNA.

Protocol

Running PCRs

1. Label 12 microfuge tubes (1.5 ml) as 1A, 2A, 3A, 4A, 5T, 6T, 7T, 8T, CA, CT, RSMA, and RSMT. Place the tubes on ice.
2. Dilute your DNA to a concentration of $50 \, ng \, \mu l^{-1}$. For example, if the concentration of your DNA is $0.6 \, \mu g \, \mu l^{-1}$ (i.e. $600 \, ng \, \mu l^{-1}$) you will need to dilute your DNA 12 times (600:50 = 12). To prepare this dilution, add 50 μl of water to the microfuge tube. Prepare a wide-bore yellow tip by cutting off 2–3 mm from the end with a razor blade. Withdraw 4 μl (50:12 = 4.1) of

water from the tube and mark the level of the solution on the tip with a marking pen. Discard the water from the tip and withdraw DNA solution to the mark. Add the DNA to the tube with water and pipette up and down several times in order to remove the viscose DNA solution from the inside of the tip.

3. Prepare the reaction stock mixture for the PCR *Alu* reaction (RSMA). Add the reagents to the RSMA tube as described in Table 4.3. Start with the addition of water and buffer. Add enzyme last. Mix all of the ingredients by pipetting up and down. Centrifuge the tube for 30 seconds and place it on ice.

4. Prepare the reaction stock mixture for the PCR STR reaction (RSMT). Add the reagents to the RSMT tube as indicated in Table 4.4. Start with the addition of water and buffer. Add enzyme last. Mix all of the ingredients by

Table 4.3 Reaction stock mixture for *Alu* amplification (RSMA)

Ingredients concentration	Add for six reactions	Amount for one reaction	Final concentration
Buffer (ten times, low Mg^{2+})	6.0 µl	1.0 µl	one times
4dNTP (2.0 mM)	6.0 µl	1.0 µl	200.0 mM
Primer 1 *Alu* F (5.0 µM)	6.0 µl	1.0 µl	0.5 µM
Primer 2 *Alu* R (5.0 µM)	6.0 µl	1.0 µl	0.5 µM
Acetamide (50%)	3.0 µl	0.5 µl	2.5%
Enzyme (5 u µl^{-1})	1.2 µl	0.2 µl	1 u/10 µl
Water	25.8 µl	4.3 µl	
Total	54.0 µl	9.0 µl	

Table 4.4 Reaction stock mixture for STR (RSMT)

Ingredient concentration	Add for six reactions	Amount for one reaction	Final concentration
Buffer (ten times, high Mg^{2+})	6.0 µl	1.0 µl	one times
4dNTP (2.0 mM)	6.0 µl	1.0 µl	200.0 mM
Primer 1 TT F (5.0 µM)	6.0 µl	1.0 µl	0.5 µM
Primer 2 TT R (5.0 µM)	6.0 µl	1.0 µl	0.5 µM
Acetamide (50%)	3.0 µl	0.5 µl	2.5%
Enzyme (5 u µl^{-1})	1.2 µl	0.2 µl	1 u 10 µl^{-1}
Water	25.8 µl	4.3 µl	
Total	54.0 µl	9.0 µl	

Table 4.5 Running PCR

	Tube number									
	1A	2A	3A	4A	CA	5T	6T	7T	8T	CT
RSMA (*Alu* mixture)	9 µl	9 µl	9 µl	9 µl	9 µl	–	–	–	–	–
RSMT (TT mixture)	–	–	–	–	–	9 µl	9 µl	9 µl	9 µl	9 µl
Your DNA	1 µl	–	–	–	1 µl	–	–	–	–	–
Partner, DNA	–	1 µl	–	–	–	–	1 µl	–	–	–
Control 1 DNA	–	–	1 µl	–	–	–	–	1 µl	–	–
Control 2 DNA	–	–	–	1 µl	–	–	–	–	1 µl	–
Water	–	–	–	–	1 µl	–	–	–	–	1 µl

pipetting up and down. Centrifuge the tube for 30 seconds and place it on ice.

5. Prepare the reactions to run a PCR for *Alu* and TT amplifications. **Keep all tubes on ice.** Add 9 µl of RSMA to the first five tubes and RSMT to the remaining tubes as indicated in Table 4.5. Tubes 1, 2, 5, and 6 will contain your and your partner's DNA, respectively. Tubes 3, 4, 7, and 8 will have control DNA. Tubes CA and CT are controls without DNA.

6. Mix each tube briefly by pipetting up and down and centrifuge for 10 seconds. Place the tubes into an ice bucket.

7. Load the reactions into 10 µl capillary tubes. Insert the open end of the capillary tube into the white silicon end of the micro-dispenser approximately 5 mm deep and draw the reaction mixture into it by slowly turning the micro-dispenser knob counterclockwise. Position the liquid in the middle of the capillary and seal by flaming the end. Only a few seconds of heating the extreme tip of the capillary tube is necessary. Check the seal by gently turning the micro-dispenser knob back and forth. If the liquid does not move in the tube, the tube is sealed. Flame seal the other end of a capillary tube. Place the sealed tube into a microfuge tube **with the end sealed last on top**.

8. Set the cycling condition as follows: D (denaturation) = 94°C for 2 seconds, A (annealing) = 55°C for 2 seconds, and E (elongation) = 72°C for 30 seconds. Start with initial denaturation at 94°C for 2 minutes and finish the reaction with an extension step at 72°C for 5 minutes. Run the reactions for 30 cycles with the S setting on 9.

9. Place the capillary tubes into the DNA thermal cycler inserting the first sealed end into the holder. Start the machine and observe the end of the capillary tube extruding from the holder carefully. If the second end is improperly sealed, liquid will rise to the top of the capillary tube when the

temperature reaches approximately 90°C. If this happens, stop the cycler and reseal the open end. Restart the cycling again.

10. After cycling is complete, remove the capillary tubes and place them into appropriately labeled 1.5 ml centrifuge tubes. Store tubes in the refrigerator until the next laboratory period.

SECOND LABORATORY PERIOD

In this laboratory period, you will analyze the results of the PCR reactions prepared in the previous laboratory period. You will use agarose gel electrophoresis for this analysis. The products of STR TT reactions will be analyzed using high resolution MetPhor™ agarose gel electrophoresis. The products of the *Alu* element amplification will be analyzed on a regular agarose gel using TBE (Tris–borate EDTA) buffer. This buffer increases the resolving power of standard agarose gels.

Safety precautions

Ethidium bromide is a mutagen and suspected carcinogen. Contact with the skin should be avoided. Wear gloves when handling ethidium bromide solution and gels containing ethidium bromide.

For safety purposes, the electrophoresis apparatus should always be placed on the laboratory bench with the positive electrode (red) facing away from the investigator, that is away from the edge of the bench. To avoid electric shock always disconnect the red (positive) lead first.

Ultraviolet (UV) light can damage the retina of the eye and cause severe sunburn. Always use safety glasses and a protective face shield to view the gel. Work in gloves and wear a long-sleeved shirt or laboratory coat when operating UV illuminators.

Protocol

High resolution agarose electrophoresis: analysis of the results of TT amplification

This procedure uses MetaPhor™ agarose that has twice the resolution capabilities of standard agarose. DNA fragments that differ by only 4 bp can be resolved using this matrix in the size range between 40 and 800 bp. MetaPhor™ agarose has approximately the resolution power of standard polyacrylamide gels.

1. Use a mini-gel-casting tray when working with this agarose (e.g. gel size 7.5 cm × 7.5 cm × 0.4 cm). Preparing smaller gels limits the cost and does not affect the resolution of DNA bands. Seal the ends of the gel-casting tray with tape. Regular labeling tape or electrical insulation tape can be used. Check that the bottom of the comb is approximately 0.5 mm above the gel bottom. To adjust this height it is most convenient to place a plastic charge card (for example MasterCard) at the bottom of the tray and adjust the comb height to a position where it is easy to remove the card from under the comb.

2. Prepare 500 ml of one times TBE buffer by adding 50 ml of a ten times TBE buffer stock solution to 450 ml of deionized water.

3. Prepare a 3.5 percent MetaPhor™ agarose mini-gel in one times TBE buffer. Place 30 ml of the buffer into a 250 ml Erlenmeyer flask and add 1.0 g agarose. Allow the powder to swell in the buffer for at least 25 minutes. Melt the agarose by heating the solution in a microwave oven at full power for 20–30 seconds at a time until the agarose is fully dissolved. The MetaPhor™ agarose is more difficult to melt than regular agarose since it "boils over" very easily. If evaporation occurs during melting, adjust the volume to 30 ml with deionized water.

4. Cool the agarose solution to approximately 60°C and add 1 μl of ethidium bromide stock solution. Slowly pour the agarose into the casting tray. Remove any air bubbles by trapping them in a 10 ml pipette.

5. Position the sample comb at approximately 1.0 cm from the edge of the gel. Let the agarose solidify for approximately 15 minutes. In order to achieve maximum resolution, after the MetaPhor™ has solidified transfer the gel to a 4°C refrigerator for 25 minutes.

6. Remove the comb with a gentle back and forth motion, taking care not to tear the gel. Remove the tape from the ends of the gel-casting tray and place the tray on the central supporting platform of the buffer chamber.

7. Add electrophoresis buffer to the buffer chamber until it reaches a level approximately 0.5–1 cm above the surface of the gel.

8. First load 14 μl of a low molecular weight standard in the first well of the gel using a P20 Pipetman.

9. Next load TT series samples into following wells. Do not skip wells between samples. Holding the capillary end tightly with two fingers 5 mm from the end gently snap off the end. Insert the open end of the capillary tube into the white silicon tip of the micro-dispenser, approximately 5 mm deep. Holding the other end of the capillary with two fingers snap it off.

10. Remove the amplified sample from the capillary tube by slowly turning the micro-dispenser knob clockwise. Do not insert the end of the capillary tube into the well; keep the end above the well. Place the tip **under** the surface of the electrophoresis buffer and **above the sample well** opening. Deliver the sample slowly, allowing it to sink to the bottom of the well. During loading it is very important not to place the tip into the well or touch the edge of the well with it. This can damage the well resulting in uneven or smeared bands.

11. Place the lid on the gel box and connect the electrodes. DNA will travel towards the positive (red) electrode positioned away from the edge of the laboratory bench. Turn on the power supply and run gel electrophoresis for 20–30 minutes at 60–80 V.

12. Turn the power off and disconnect the leads from the power supply. To avoid electric shock always disconnect the red (positive) lead first.

13. Remove the gel tray from the electrophoresis chamber and place it on a UV illuminator. Photograph the gel using Polaroid film 667 and a Polaroid camera at speed 1 and F stop 8 or use a computer imaging system for recording the results.

High resolution agarose electrophoresis: analysis of the results of Alu *amplification*

A gel prepared in TBE buffer will be used for this analysis. Agarose gel prepared in TBE buffer has lower porosity than TAE (Tris–acetate EDTA) buffer-prepared gels. The DNA mobility in this buffer is approximately two times slower than in TAE buffer, but the DNA bands are sharper. This gel has the best resolving power for DNA fragments of less than 1,000 bp.

1. Prepare a 1.2 percent agarose mini-gel in one times TBE buffer. Use a mini-gel-casting tray (e.g. gel size 7.5 cm × 7.5 cm × 0.4 cm). Seal the ends of the gel-casting tray with tape. Regular labeling tape or electrical insulation tape can be used. Check that the bottom of the comb is approximately 0.5 mm above the gel bottom as described in step 1 of the previous procedure.
2. Use 30 ml of agarose solution. Weigh 360 mg of agarose and add it to 30 ml of one times TBE buffer. Melt the agarose in a microwave oven, adjust the volume to 30 ml with water, and add 1.0 µl of ethidium bromide. Pour immediately into the gel form. Let it solidify for 20–30 minutes.
3. First load 14 µl of a low molecular weight standard in the first well of the gel using a P20 Pipetman.
4. Next load *Alu* series samples into consecutive wells. Do not skip wells between samples. Holding the capillary end tightly with two fingers gently snap it off 5 mm from the end. Insert this end of the capillary tube into the white silicon tip of the micro-dispenser, approximately 5 mm deep. Holding the other end of the capillary with two fingers snap it off.
5. Remove the amplified sample from the capillary tube by slowly turning the micro-dispenser knob clockwise. Do not insert the end of capillary tube into the well and keep the end above the well. Place the tip **under** the surface of the electrophoresis buffer and **above the sample well** opening. Deliver the sample slowly, allowing it to sink to the bottom of the well. During loading it is very important not to place the tip into the well or touch the edge of the well with it. This can damage the well resulting in uneven or smeared bands.
6. Place the lid on the gel box and connect the electrodes. DNA will travel towards the positive (red) electrode positioned away from the edge of the laboratory bench. Turn on the power supply and run gel electrophoresis for 30 minutes at 110 V.
7. Turn the power off and disconnect the leads from the power supply. To avoid electric shock always disconnect the red (positive) lead first.
8. Remove the gel tray from the electrophoresis chamber and place it on a

UV illuminator. Photograph the gel using Polaroid film 667 at speed 1 and F stop 8 or use a computer imaging system to record the results.

Data analysis

Examine your *Alu* sample gel and determine to which *Alu* haplotype you belong. The presence of two bands indicates that you are heterozygote. A single band indicates you are a homozygote with or without the *Alu* insertion. Look for examples in Fig. 4.3.

Examine your TT sample gel. Determine your haplotype for STR elements. Consult Fig. 4.4 for examples of band locations and their sizes.

Using the data collected above, determine haplotype *Alu* and STR for both of your chromosomes. For example, individual 1 is heterozygotic for *Alu* insertion (well 1 in Fig. 4.3) and homozygotic for STR 85 (well 1 in Fig. 4.3). The haplotypes are as follows: *Alu*(+) STR 85 and *Alu*(−) STR 85. Inspection of Table 4.2 will indicate that this individual cannot belong to New World or Pacific Island populations.

In some instances exact determination of each haplotype will not be possible using the data collected. This situation occurs if you are heterozygotic for both loci. For example, individual 2 in Fig. 4.3 is heterozygotic for *Alu* insertion and also heterozygotic for STR 110 and STR 90 (Fig. 4.4). This person can have the following haplotypes.

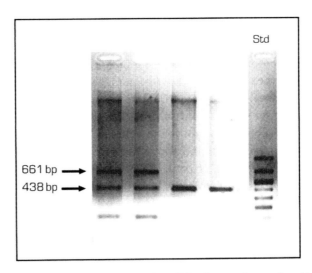

Figure 4.3 Agarose gel electrophoresis of amplification products using *Alu* primers. Wells 1 and 2 contain the DNA of heterozygotes containing chromosome with and without *Alu*. Wells 3 and 4 contain DNA from homozygotes with *Alu* deletion. Arrows indicate the position and size of haplotypes: *Alu* insertion (661 bp) and *Alu* deletion (438 bp), respectively.

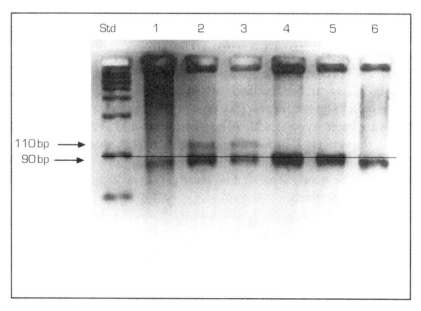

Figure 4.4 Amplification of human genomic DNA using STR primers. Wells 2 and 3 contain the DNA of heterozygotes with ten repeats (110 bp) and six repeats (90 bp). Wells 1 and 6 contain the DNA of homozygotes with five repeats (85 bp). Wells 4 and 5 contain the DNA of homozygotes with six repeats (90 bp). The line marks the position of a DNA standard fragment that is 100 bp long.

1. *Alu*(+) STR 90.
2. *Alu*(−) STR 90.
3. *Alu*(+) STR 110.
4. *Alu*(−) STR 110.

Inspection of Table 4.2 indicates that haplotype *Alu*(−) STR 110 is unlikely because it appears only in the European population. The most probable haplotypes for this individual are *Alu*(−) STR 90 and *Alu*(+) STR 110. This is because STR 90 allele is most frequently associated with *Alu*(−) and STR 110 allele is most frequently associated with *Alu*(+) in whole world populations, respectively.

Compile the data from the entire class and calculate the class δ value for the *Alu*(−) haplotype associated with STR 85, STR 90, and STR 115 ($\delta_{-,STR\,85}$, $\delta_{-,STR\,90}$, and $\delta_{-,STR\,115}$) as described in the Introduction. Compare the calculated values to the δ values of world populations (Table 4.1).

Expected results

Figure 4.3 presents the amplification results of the *Alu* element polymorphism. All three haplotypes are clearly identified. Individuals 1 and 2 are heterozygotes with one chromosome carrying an *Alu* insertion (661 bp) and the second having an *Alu* deletion (438 bp). Individuals 3 and 4 are homozygotic with respect to *Alu* deletion.

Figure 4.4 presents the amplification results for STRs. The three most common alleles can be identified with ease. The first and sixth individuals have only an 85 bp long repeat (five repeats) and the second and third individuals are heterozygotes with 90 bp (six repeats) and 110 bp (ten repeats) repeats, respectively. The fourth and the fifth individuals are homozygotes with a 90 bp long repeat. The other less common haplotypes 95 bp and 80 bp long cannot be clearly separated on these gels from haplotypes 90 bp and 85 bp long, respectively.

References

Cann, R., Stoneking, M., and Wilson, A.C. (1987) Mitochondrial DNA and human evolution. *Nature*, **325**, 31–6.

Casalotti, R., Simoni, L., Belledi, M., and Barbujani, G. (1999) Y-chromosome polymorphism and the origin of the European gene pool. *Proc. R. Soc. Lond.* B, **266**, 1959–65.

Innis, M.A., Gelfand, D.H., Sninsky, J.J., and White, T.J. (eds) (1990) *PCR Protocols: A Guide to Methods and Applications*. Academic Press, San Diego, CA.

Richards, M., Corte-Real, H., Forster, P., Macaulay, V., Wilkinson-Herbots, H., Demaine, A. et al. (1996) Palaeolithic and Neolithic lineages in the mitochondrial gene pool. *Am. J. Human. Genet.*, **59**, 185–203.

Sardelli, A.D. (1993) Plateau effect: understanding PCR limitations. *Amplification*, **9**, 1–5.

Seielstad, M., Bakele, E., Ibrahim, M., Toure, A., and Traore, M. (1999) A view of modern human origin from Y chromosome microsatellite variation. *Genome Res.*, **9**, 558–67.

Semino, O., Passarino, G., Brega, A., Fellous, M., and Santachiara-Benerecetti, A.S. (1996) A view of the Neolithic demic diffusion in Europe through two Y chromosome-specific markers. *Am. J. Genet.*, **59**, 964–8.

Tishkoff, S.A., Dietzsch, E., Speed, W., Paksitis, A., Kidd, J., Cheung, K. et al. (1996) Global patterns of linkage disequilibrium at the CD4 locus and modern human origins. *Science*, **271**, 1380–7.

Wittwer, C.T. and Garling, D.J. (1991) Rapid cycle DNA amplification: time and temperature optimization. *BioTechniques*, **10**, 76–83.

Wittwer, C.T., Reed, G.B., and Ririe, K.M. (1994) Rapid cycle DNA amplification. In *The Polymerase Chain Reaction*, K.B. Mullis (ed.), pp. 174–81. Birkhauser, New York.

CHAPTER 5

DNA Sequencing

Introduction

The goal of this experiment is to sequence human DNA using the same procedures employed in large sequencing projects. The process of DNA sequencing consists of three basic tasks. The first task is the generation of the individual fragments to be sequenced. The second involves running sequencing reactions and the third electrophoresis and compilation of the data. For this exercise you will use DNA isolated from your cheek cells. DNA sequencing will be carried out using the technique of random or shotgun sequencing. This requires preparation of a sequencing library. This technique was the main technique used in the human genome project, which encompasses sequencing approximately 30 million short DNA fragments (International Human Genome Sequencing Consortium, 2001; Vender et al., 2001). You will sequence only three such DNA fragments prepared from your genomic DNA. However, you will perform all steps for preparing a sequencing library and sequence these fragments. This technique was utilized in sequencing the entire human genome.

Sequencing DNA using the shotgun or random fragment sequencing method involves the following steps.

1. Fragmentation of DNA into random short fragments using a nebulization procedure. Experiment 1 describes this procedure.
2. Repairing the ends of the sheared DNA fragments in order to create blunt-ended DNA fragments. Experiment 2 describes this technique.
3. Blunt-end ligation of repaired fragments into sequencing vector (pUC 18 plasmid). The conditions of ligation are selected to permit ligation of only a single fragment into one vector molecule. The ligation procedure is the subject of experiment 3.
4. Transforming bacteria with chimeric plasmids using electroporation. This procedure will be described in experiment 4.
5. Isolation of the plasmid from a single bacterial colony and sequencing insert DNA using the cycle sequencing reaction. Samples are prepared

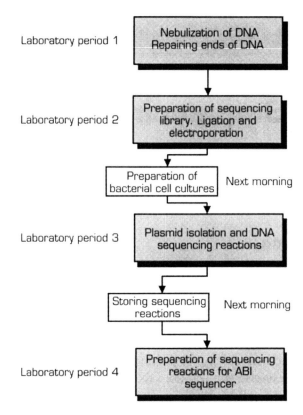

Figure 5.1 Schematic outline of the procedures used in a DNA sequencing laboratory.

for sequencing and sequenced in an ABI Sequencer. These procedures will be carried out in experiments 5–7.

The entire experiment will be done during four laboratory periods. Figure 5.1 presents the overall timetable for these experiments.

Background

Determination of a DNA sequence is the only method in biological science that generates data that are not biased by previous assumptions, hypotheses, or experimental designs. It is therefore not surprising that DNA sequencing has revealed many unexpected facts concerning gene structure, the regulation of gene expression, and organization of genomes, as well as discovering new genes never seen before. Advances in large-scale sequencing have also brought about new scientific disciplines, for example genomics and functional genomics, that are devoted to analyzing whole genomes and their function (Hieter and Boguski, 1997).

DNA sequencing methods

All of the present methods of DNA sequencing are based on the anchored-end principle. In these methods, one end of the sequenced DNA molecule remains unchanged (anchored) while the other is generated in a base-dependent way. This creates sets of DNA molecules of various lengths having one end common and the other end terminating at a specific base. Separation of these molecules according to length generates the base sequence of the fragment. The separation is usually accomplished by electrophoresis using a matrix capable of distinguishing between two DNA molecules differing only by a single nucleotide.

There are two methods that generate DNA molecules of different lengths in a base-specific manner. The first method, which was introduced by Maxam and Gilbert (1977), uses base-specific chemical cleavage of the DNA fragment. The second method uses the enzymatic synthesis of DNA fragments (Sanger et al., 1977).

In the Maxam and Gilbert (1977) method of chemical degradation, one end of the DNA is labeled and four separately run base modification reactions are performed. One of the bases (A, C, G, or T) is modified in each reaction. The DNA backbone is then cleaved at each modified residue, resulting in a "nested set" of fragments all labeled at one end and terminating at the location of a specific modified base. Fragments are then separated using gel electrophoresis. This method of DNA sequencing is rarely used at the present time.

The enzymatic method of DNA sequencing utilizes properties of DNA polymerase for implementing the anchored-end principle. DNA polymerase can synthesize a complementary strand from a single-stranded template. Initiation of this synthesis is dependent on the presence of a primer with a 3′-hydroxyl group. In addition, procaryotic DNA polymerases are able to incorporate a dideoxynucleotide instead of deoxynucleotide into the growing DNA chain. Thus, all synthesized DNA molecules will share an identical 5′-end and the 5′-end of the primer and the 3′-end will terminate at specific bases by incorporation of the substrate lacking a 3′-hydroxyl residue.

Four primer extension reactions are initiated using the same primer. Each reaction contains all four usual 2′-deoxynucleotides (dNTPs), but only one of the four 2′,3′-dideoxynucleotides (ddNTPs). By carefully controlling the ratio between dNTP and ddNTP in each of the four reactions, incorporation of the dideoxy nucleotide and, hence, chain termination is random. The end result is the generation of a set of DNA fragments of different lengths, each terminated at the 3′-end at a specific base. These fragments are then separated using electrophoresis.

The length of a DNA molecule that can be sequenced using the anchored-end principle depends not on the ability to create different sized "anchored" fragments, but on the resolving power of the matrix used for separating these fragments. The resolving power of matrices presently used permits separation of DNA fragments up to approximately 1,000 bases, setting the limit on the length of DNA fragment that can be sequenced in a single reaction.

Sequencing strategies

Sequence determination of a DNA fragment smaller than 1,000 bases is relatively simple. It requires cloning the fragment into an appropriate single- or double-stranded DNA sequencing vector, sequencing the fragment in a single sequencing reaction, and running electrophoresis to separate these fragments. In order to sequence a large DNA molecule it is necessary to subdivide this fragment into smaller fragments of approximately 1,000 bases long. Each small fragment is sequenced separately in a single sequencing reaction and the sequence of the whole fragment is assembled. The way in which smaller fragments are generated from the large DNA fragment and then assembled into the sequence of the whole fragment is referred to as the **sequencing strategy**.

There are two groups of sequencing strategy: directed strategies, in which specific starting points are used for sequencing and random strategies, in which the starting points for sequencing are random (Hunkapiller et al., 1991). Directed strategies permit the direct sequencing of a large region by generating small fragments, the position of which in the whole molecule is known. Usually sequencing proceeds successively from one end of the large fragment to the other. Random strategies (shotgun strategies) involve sequencing of fragments generated by random shearing of a large piece of DNA or random insertions of a universal primer site into target DNA using a transposon.

Primer walking or primer-directed strategy is the most frequently used directed strategy and can be used for sequencing DNA fragments of 10,000 bases or longer. The entire fragment is cloned into a sequencing vector and the initial sequence data are obtained using a vector-based universal primer. The sequences obtained are used for synthesizing a new primer that hybridizes near the 3′-end of the newly elucidated sequence. This primer is used to sequence the next DNA fragment. The cycle of sequencing and primer synthesis is repeated until the whole fragment is sequenced. The technique is uniquely suited to the dideoxy DNA sequencing method and bypasses the need for subcloning smaller pieces of DNA.

In the random sequencing strategy randomly generated fragments are subcloned into a sequencing vector, forming a **sequencing library** of the fragment. Next, the fragments are randomly chosen for sequencing from this library. Identifying overlaps between small fragments and arranging them into the most probable order assembles the sequence of the original piece of DNA. The number of clones necessary for assembling the whole fragment (**subclone coverage**) and the amount of raw data (**sequence coverage**) are directly proportional to the size of the target DNA. The relationship between these parameters (Deininger, 1983) is given by

$$S = 1 - \left(1 - \frac{i}{L}\right)^n \tag{5.1}$$

where S is the fraction of the fragment sequenced, i is the average number of bases read per clone, L is the length of the whole fragment (bp), and n is the number of clones of length i sequenced. Since the average length of sequence read per subclone is usually approximately 400 bases, to assemble 95 percent of the sequence requires $n = 3L/400$ subclones to be sequenced giving a sequence redundancy of 3.0. For example, for a 40,000 bp fragment sequencing 300 clones of 400 bases gives 95 percent of the whole sequence $(0.95 = 1 - (1 - 400/40,000)^n)$. For both strand coverage of the same fragment the number of clones should be doubled so n will be equal to $n = 6L/400$ giving a redundancy of 6.0 and sequencing of 99.7 percent of the whole fragment.

Random strategy using the chain termination method of DNA sequencing is a method of choice for sequencing large DNA fragments or whole genomes. Directed strategies, such as primer walking and deletion strategy, are used for sequencing shorter DNA fragments or for filling gaps between contigs.

References

Deininger, P.L. (1983) Random subcloning of sonicated DNA: application to shotgun DNA sequence analysis. *Anal. Biochem.*, **129**, 216–23.
Hieter, P. and Boguski, M. (1997) Functional genomics: it's all how you read it. *Science*, **278**, 601–2.
Hunkapiller, T., Kaiser, R.J., Koop, B.F., and Hood, L. (1991) Large-scale and automated DNA sequence determination. *Science*, **254**, 59–67.
International Human Genome Sequencing Consortium (2001) Initial sequencing and analysis of the human genome. *Nature*, **409**, 860–921.
Maxam, A. and Gilbert, W. (1977) A new method for sequencing DNA. *Proc. Natl Acad. Sci. USA*, **74**, 560–4.
Sanger, F., Nicklen, S., and Coulson, A. (1977) DNA sequencing with chain-terminating inhibitors. *Proc. Natl Acad. Sci. USA*, **74**, 4298–5467.
Vender, J.C. et al. (2001) The sequence of human genome. *Science*, **291**, 1305–51.

FIRST LABORATORY PERIOD

In this laboratory period you will begin to prepare a sequencing library of your DNA. To sequence DNA using the shotgun or random fragment sequencing method, DNA should be first sheared into random short fragments. The fragmentation of DNA will be accomplished using a nebulization process. Next, you will "repair" the ends of the sheared DNA fragments in order to create blunt-ended fragments that can be cloned into sequencing plasmids.

Experiment 1: nebulization shearing of DNA

Introduction

To prepare a sequencing library, DNA must be fragmented into small fragments. In this experiment you will prepare DNA fragments using a novel DNA shearing method. You will shear your DNA that you prepared in a previous laboratory exercise.

A truly useful method for producing random subclones should have the following properties. First, it should produce truly random DNA fragments, i.e. shearing should be sequence independent. Second, it should be reproducible at any time and with any DNA. In order to achieve this, shearing should be reached in a steady-state manner; i.e. shearing to a particular size should not be dependent on the time of application of the shearing agent. Third, the method should allow the generation of DNA fragments in a size range of 500–2,000 bp. Fourth, the method should be efficient and the majority of the DNA treated should be converted into the desired size fragments. The method that fulfills all of these requirements uses dynamic shearing of DNA molecules in the process of nebulization (Surzycki, 1990).

Background

Application of a random strategy for sequencing large pieces of DNA requires the preparation of a sequencing library that contains DNA fragments of approximately 1,000–2,000 bp long that are randomly generated from the original DNA.

There are four methods used for generating random DNA fragments. The first method employs partial restriction enzyme digestion. The second involves fragmentation of DNA by DNase I in the presence of Mn^{2+}, the third relies on sonication to physically break the DNA, and the fourth uses a nebulization process for shearing DNA. However, there are a number of major disadvantages in the use of the first three methods.

The major drawback of the first method, the use of restriction enzymes, is the lack of randomness in the clone bank because the distribution of restriction sites along the DNA is not random. This necessitates using a number of different restriction enzymes for the preparation of sequencing banks, which is a laborious and time-consuming process. This method also requires performing a number of carefully controlled restriction enzyme reactions that are difficult to reproduce with different enzymes and DNA preparations.

The use of DNase surmounts some of the difficulties in the first method because there is very little DNA sequence specificity in DNase cleavage. However, even to a larger extent than in the first method, the application of DNase in generating random fragments is difficult to reproduce and requires numerous test reactions. This is wasteful and necessitates having large amounts of starting material.

The advantage of the third method is that it is easier to reproduce and control than either of the enzymatic methods. However, its application requires large amounts of starting material because only a small portion of sheared DNA molecules are the required size. This method also involves laborious calibration of the sonicator and rigorous timing for subsequent treatments. Moreover, it has been shown that sonication shears AT-rich sequences preferentially and, thus, does not create truly random sequencing libraries. This is particularly evident if the DNA to be sheared is composed of long AT- and GC-rich stretches.

The nebulization method avoids all of these difficulties and is now used for preparation of random sequencing libraries (Surzycki, 1990). The method works on the following principle. In the process of nebulization or reducing liquid to a fine spray, small liquid droplets of uniform size are produced. In the process of droplet formation the liquid being nebulized flows from the liquid surface to the forming bubble. This creates a transient flow between the liquid surface and the droplet through a connecting small capillary. The diameter of this capillary channel is approximately half the diameter of the forming droplet and can be adjusted as desired by controlling the size of the droplets. The velocity of flow of the liquid in the capillary is not constant across the capillary due to frictional resistance between adjacent layers of flowing liquid. The velocity gradient generated causes liquid in the center to flow faster than liquid in the outer layers creating flow that is called **laminar flow**. Because the velocity of flow is not constant across the capillary tube, the DNA molecule that finds itself in two adjacent flow layers will have its ends moving at two different velocities. This results in stretching and rotating of the molecule until it is positioned in a single laminar layer or broken at the point where the stretching force is maximal.

For large, ridged, linear molecules, such as DNA, the stretching force is the greatest at the middle of the molecule. Consequently, the molecule will have the greatest probability of breaking in half. The nebulization of DNA will result in the breakage of each molecule almost exactly in half in the repeated process of bubble formation. This will continue until the molecule reaches a small enough size that it cannot be positioned across two laminar flow layers. The final size of the broken molecules will depend only on the size of the droplet formed, i.e. the size of the capillary nebulization channel, but **not on the time of nebulization**.

The nebulization process permits the regulation of the size of the DNA fragments generated. The formation of droplets can be considered at two levels of nebulization: the primary nebulization process that results in the formation of primary droplets and the secondary nebulization process resulting in the formation of secondary droplets by the shattering of the primary droplets on the surface of the nebulization sphere. Droplets are formed at both sides by laminar flow of the liquid, generating force that is formally described by the equation of liquid capillary flow. Accordingly, this force is directly proportional to the gas pressure applied and the viscosity of the liquid and inversely proportional to the size of the droplets. Consequently, the smaller the droplets and the higher viscosity and greater gas pressure applied, the larger the force that creates smaller DNA fragments.

The droplet diameter that is necessary for shearing DNA molecules is in the order of $0.1-2.0\mu$. This droplet size is created only at the site of secondary nebulization. According to the equation describing the droplet size formed during the secondary nebulization process the size of the droplet depends on the velocity of the primary droplets, the absolute viscosity of the gas used, and the diameter of the droplets generated in the primary nebulization process.

In order to decrease the diameter of primary droplets, DNA is nebulized in a solution of 25 percent glycerol. Moreover, nitrogen or argon is used for achieving droplets with the correct diameter in secondary nebulization. Because the absolute viscosity of these gases is high and the viscosity of nitrogen is very close to the absolute viscosity of argon these gases can be used interchangeably. Use of air, that is a mixture of gases of different absolute viscosity, generates a different size droplet that consequently leads to a broad distribution of fragment sizes.

Safety precautions

Ethidium bromide is a mutagen and suspected carcinogen. Contact with the skin should be avoided. Wear gloves when handling ethidium bromide solution and gels containing ethidium bromide.

For safety purposes, the electrophoresis apparatus should always be placed on the laboratory bench with the positive electrode (red) facing away from the investigator, that is away from the edge of the bench. To avoid electric shock always disconnect the red (positive) lead first.

Ultraviolet (UV) light can damage the retina of the eye and cause severe sunburn. Always use safety glasses and a protective face shield to view the gel. Work in gloves and wear a long-sleeved shirt or laboratory coat when operating UV illuminators.

Technical tips

Between 1 and 5 μg of DNA should be nebulized in order to create a complete sequencing library. The purity and size of DNA is not critical for successful nebulization. A very good library can be prepared using DNA as small as 30–50 kbp. The nebulization volume used here is 1 ml. The breathing hole of the inhalator should be closed in order to limit DNA lost with escaping mist. A cut end of a 15 ml plastic centrifuge tube can be used (Surzycki, 1990). The opening can also be covered with a QS-T cap (Roe and Crabtree, 1995; Hengen, 1997).

Nitrogen or argon gas should be used for nebulization in order to obtain a library of uniform fragment size. Substituting any other gas or compressed air is possible, but will result in a broader size distribution of fragments. If a gas pressure regulator is used, the pressure should be set to 30 psi. This pressure should generate a DNA fragment of approximately 1,000 bp. If larger fragments are desired lower pressure should be used. For example, to obtain a DNA fragment of 3,000 bp nebulization should be carried out at 10 psi (Surzycki, 2000).

The time of nebulization when using a lower gas pressure should not be changed. This is because the time of nebulization depends only on the volume of nebulized liquid.

The minimum amount of liquid that can be used in an inhalator type nebulizer is 1 ml. Size fractionation of nebulized DNA by gel electrophoresis is not necessary. More than 85 percent of DNA fragments are of the desired size because the nebulization process cannot generate DNA fragments smaller than that set by the droplet size. Larger fragments can be generated by incomplete nebulization, but even when they are present their cloning efficiency is low and these fragments are not present in the sequencing library.

Incomplete nebulization can result from a defective nebulizer, a volume that exceeds 1 ml of nebulized DNA, or low gas pressure. It is important to realize **that circular and supercoiled DNA** are not sheared by nebulization.

Running gel electrophoresis of nebulized samples is optional. Moreover, only one or two gels need to be prepared for an entire class since each student will have a single nebulized sample of her or his DNA. Even if the sheared DNA is not visible, as shown in Fig. 5.2, students should continue processing their sample in order to prepare their sequencing library. A sequencing library can be prepared successfully using a very small amount of sheared DNA that cannot be visualized on ethidium bromide-stained gels.

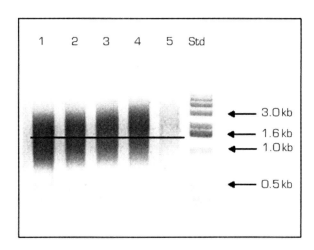

Figure 5.2 Agarose gel electrophoresis of human genomic DNA nebulized for various times using nitrogen at 30 psi. Lane 1 is nebulization for 90 seconds, lane 2 is nebulization for 75 seconds, lane 3 is nebulization for 55 seconds, lane 4 is nebulization for 35 seconds, and lane 5 is nebulization for 15 seconds. The STD lane is a molecular weight standard ladder of 1 kb. The line indicates the position of 1.5 kb DNA.

Protocol

1. Label two sterile 1.5 ml microfuge tubes SP and 0 and place them on ice. Add 150 µl of TE buffer to the SP tube. Next you will add 2–10 µg of your DNA to this tube. To do so you need first to calculate how many microliters of your DNA you should add. **Calculation example:** if the concentration of your DNA is 0.5 µg µl^{-1}, in order to obtain 2 µg of DNA you need to take 4 µl of your DNA (2/0.5 = 4.0 µl).

2. Add the calculated amount of DNA to the SP tube and mix by pipetting up and down.

3. You will be given a tube with 0.85 ml of TE buffer (pH 7.5) containing 25 percent glycerol. This is the solution for nebulization. Add the DNA prepared in the SP tube to it.

4. Place 20 µl of the DNA mixture into tube 0. This will constitute the control, non-nebulized DNA.

5. Transfer the remaining DNA solution into the bottom of the nebulization apparatus using a P1000 Pipetman. Close the nebulizer top and insert the "closing valve" into the breathing hole of the inhalator.

6. Attach the instrument to the nitrogen gas tank pressure regulator using the plastic tubing provided. **Note:** the nebulizer can be attached to a laboratory-compressed air line. To obtain the desired air pressure, adjust the laboratory air outlet to approximately one-third open.

7. Set the gas pressure regulator to 30 psi (2.0 atm). Open the gas pressure regulator valve and nebulize in 15 second intervals for a total time of

75 seconds. You will see a little mist escaping under the "closing valve." This indicates that the nebulizer is operating properly. After each 15 second interval tap the nebulization apparatus slightly on the laboratory bench in order to collect all of the droplets at the bottom of the reservoir. **Note:** if you are using laboratory-compressed air stop nebulization by removing **the hose from the air line and restart it by reconnecting it**. Do not close the air valve. This will assure constant air pressure during the entire nebulization process.

8. Transfer liquid from the nebulizer to a 1.5 ml microfuge tube. Place 15 µl of the nebulized DNA into a separate tube and add 2 µl of stop solution. Mix well by pipetting up and down. Prepare a mini-gel using a casting tray no larger than 7.5 cm × 7.5 cm and a thin gel (0.2 cm). Seal the ends of the gel-casting tray with tape. Regular labeling tape or electrical insulation tape can be used. Use a mini-gel well-casting comb with wells of 0.2–0.5 cm long and 1 mm (or less) wide. Check that the bottom of the comb is approximately 0.5 mm above the gel bottom. To adjust this height it is most convenient to place a plastic charge card (for example MasterCard) at the bottom of the tray and adjust the comb height to a position where it is easy to remove the card from under the comb. **Note:** a single gel can accommodate 12 samples including a standard. Thus, the samples of an entire class can be analyzed on one or two agarose gels.

9. Prepare 500 ml of one times TAE (Tris–acetate EDTA) buffer by adding 10 ml of a 50 times TAE stock solution to a final volume of 500 ml of deionized water.

10. Prepare 1 percent gel. Place 20 ml of the buffer into a 150 ml flask and add 200 mg of agarose. Melt the agarose by heating the solution in a microwave oven at full power for approximately 2 minutes. Carefully swirl the agarose solution in order to ensure that the agarose is dissolved, i.e. no agarose particles are visible. If evaporation occurs during melting adjust the volume to 20 ml with deionized water.

11. Cool the agarose solution to approximately 60°C and add 1 µl of ethidium bromide stock solution. Slowly pour the agarose into the gel-casting tray. Remove any air bubbles by trapping them in a 10 ml pipette.

12. Position the comb approximately 1.5 cm from the edge of the gel. Let the agarose solidify for approximately 20–30 minutes. After the agarose has solidified, remove the comb with a gentle back and forth motion, taking care not to tear the gel.

13. Remove the tape from the ends of the gel-casting tray and place the tray on the central supporting platform of the gel box. For safety purposes, the electrophoresis apparatus should be always placed on the laboratory bench with the positive electrode (red) facing away from the investigator, that is away from the edge of the bench.

14. Add electrophoresis buffer to the buffer chamber until it reaches a level of 0.5–1 cm above the surface of the gel.

15. Load 8 µl of 1 kb standards into the first well. Load samples prepared in step 8 onto the remaining wells and run the gel at 80–90 V for 30 minutes.
16. Photograph the gel. Use a setting of 1 second at F8 with Polaroid 667 film. One can also use a computer-imaging system for recording the results.

Expected results

Figure 5.2 presents gel electrophoresis of DNA nebulized for various times using nitrogen at 30 psi. Increasing the time of nebulization to longer than 75 seconds does not generate smaller fragments (wells 2 and 3). This is because the time of nebulization depends on the volume of the liquid to be nebulized. The size of the DNA fragments generated depends only on the gas pressure applied and, consequently, on the droplet size.

Experiment 2: repair of the ends of sheared DNA

Introduction

DNA molecules that have been sheared by the process of nebulization have three types of ends. Approximately 12 percent of the molecules have blunt ends at both ends of the molecules. The remainder of the DNA molecules contain various lengths of unevenly sheared ends with single-stranded DNA chains protruding for several nucleotides at the 5′- or 3′-strands. These ends cannot be ligated into blunt-ended, open vector molecules. To achieve high efficiency in library preparations, these protruding ends should be "repaired." To repair these ends, you will use T4 DNA polymerase. T4 DNA polymerase is capable of filling in recessed 3′-ends of DNA molecules in the presence of substrate by synthesizing short pieces of DNA. In addition to repair activity, T4 DNA polymerase possesses 3′-exonuclease activity that will remove protruding, single-stranded 3′-ends from sheared DNA, creating in the process a blunt-ended DNA fragment. DNA polymerase repaired ends are devoid of the 5′-phosphates that are needed to clone the fragment into a dephosphorylated vector. We will use T4 polynucleotide kinase to phosphorylate the 5′-ends lacking phosphate. The 5′-phosphates are necessary for ligase activity, in order to ligate the fragments into the sequencing vector.

Figure 5.3 presents a schematic representation of conversion of 5′- or 3′-recessed DNA fragment ends in the process of end repairing.

Safety precautions

PCI reagent is an equal mixture of phenol and chloroform:isoamyl alcohol (CIA). The reagent can be rapidly absorbed by and is highly corrosive to the

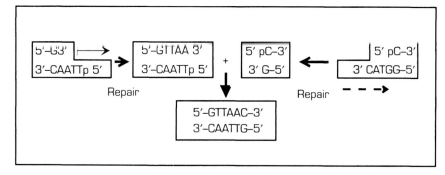

Figure 5.3 Schematic representation of a repair reaction and blunt-end ligation. The recessed 3'- and 5'-ends are converted to blunt ends by the action of T4 DNA polymerase.

skin. It initially produces a white softened area, followed by severe burns. Because of the local anesthetic properties of phenol, skin burns may not be felt until there has been serious damage. Gloves should be worn when working with this reagent. Because some brands of gloves are soluble or permeable to phenol, they should be tested before use. If PCI is spilled on the skin, flush off immediately with a large amount of water and treat with a 70 percent solution of PEG 4000 in water. Phase Lock Gel™ (PLG) tubes containing PCI should be collected in a tightly closed glass receptacle and stored in a chemical hood until proper disposal.

Technical tips

It is important to use T4 DNA polymerase rather than Klenow fragment enzyme in this experiment. The T4 DNA polymerase should be used instead of Klenow enzyme because of the following.
1. It has a much higher exonuclease activity than that of Klenow enzyme. This assures rapid removal of protruding 3'-ends.
2. It is active in many different buffers and the end repair reaction could be done simultaneously with phosphorylation of 5'-ends by polynucleotide kinase using the kinase-specific buffer.
3. The enzyme is inexpensive and much more stable than Klenow enzyme.

It is important to know the DNA concentration in your repair reaction because this is the concentration of blunt-ended DNA fragments. It is necessary to know this concentration in order to calculate the correct insert to plasmid ratio in the ligation reaction. The method described for removing enzyme from the repair reaction will recover nearly 90 percent of the DNA. It is not possible to measure the concentration of DNA fragments using a standard spectroscopic method of absorption at 260 nm. This is because the spectroscopic method requires a DNA concentration of at least 5 µg ml^{-1} ($A_{260} = 0.1$) and a minimum sample volume of 100 µl. DNA

Table 5.1 DNA end-repair reaction

Ingredients	Add to reaction	Final concentration
Ten times polynucleotide kinase buffer	5.0 μl	one times
Ten times dNTP solution	5.0 μl	200.0 μM
Nebulized DNA fragments (2–10 ng μl^{-1})	30.0 μl	0.06–0.3 μg
ATP 10 mM	5.0 μl	1.0 mM
T4 polynucleotide kinase (10 u μl^{-1})	1.0 μl	10.0 units
T4 DNA polymerase (3 u μl^{-1})	1.0 μl	3.0 units
Water	To 50.0 μl	

fragments prepared for cloning can rarely be obtained at this concentration and volume.

The most important component of the repair reaction is ATP upon which the phosphorylation of 5′-ends of DNA and, consequently, successful ligation critically depends. The stock solution should be freshly prepared and neutralized.

It is critical to remove polynucleotide kinase after the repair reaction because even a small amount of this enzyme present during ligation will phosphorylate 5′-ends of dephosphorylated plasmid. This will result in religation of plasmid molecules and, consequently, very low frequency of plasmids with inserts.

To remove polynucleotide kinase, it is best to use phenol–CIA extraction and PLG microfuge tubes. The PLG tubes contain a proprietary compound that, when centrifuged, migrates to form a tight barrier between organic and aqueous phases. The interphase material is trapped in and below this barrier allowing the complete and easy collection of the entire aqueous phase without contaminating it with organic solvents. The PLG barrier also offers increased protection from exposure to organic solvents.

Alternatively, polynucleotide kinase can be heat inactivated at 65°C for 20 minutes. Inactivation of T4 polymerase is not necessary since its presence will not interfere with the ligation reaction.

Protocol

1. Place a 1.5 ml microfuge tube on ice and label it RE. You will be given two additional tubes, one containing ten times dNTP mixture and another with ten times polynucleotide kinase buffer. The DNA that will be repaired is your nebulized DNA.

2. The total reaction volume is 50 μl. Assemble the reaction mixture in the RE tube using the ingredients indicated in Table 5.1.

3. First add the required amount of water. Next add buffer, ten times dNTP solution, and DNA. Mix by pipetting up and down.

4. Start the reaction by the addition of both enzymes. Mix by pipetting up and down. Avoid creating air bubbles in the process. Centrifuge for 5–10 seconds in order to collect the liquid at the bottom of the tube.

5. Incubate at room temperature for 25 minutes.

6. Stop reactions by the addition of 1 µl 0.5 M EDTA and 50 µl of TE buffer. Mix well by pipetting up and down.

7. Add 100 µl of PCI solution and mix by inverting the tube several times to form an emulsion.

8. Centrifuge the empty PLG tube in a microfuge for 30 seconds at 10,000 r.p.m. in order to pellet the gel. Orient the tube in the centrifuge rotor with the lid connector pointing away from the center of rotation. Measure the time of centrifugation from the moment of **starting the microfuge**.

9. Transfer 100 µl of the repair reaction mixture prepared in step 7 into the PLG tube.

10. Centrifuge the PLG tube at 10,000 r.p.m. for exactly 30 seconds from starting the centrifuge. **Be sure to orient the tube in the centrifuge rotor with the lid connector pointing away from the center of rotation.** After centrifugation, the organic phase at the bottom of the tube will be separated from the aqueous phase at the top of the tube by the PLG barrier.

11. Remove the tube from the centrifuge and add 100 µl of CIA solution to the aqueous phase. Mix by repeated inversion to form an emulsion. Do not vortex or allow the bottom organic phase to mix with the upper aqueous phase.

12. Centrifuge the PLG tube at 10,000 r.p.m. for exactly 30 seconds from starting the centrifuge. **Be sure to orient the tube in the centrifuge rotor with the lid connector pointing away from the center of rotation.**

13. Transfer the aqueous phase, collected from the above the gel, to a fresh 1.5 ml microfuge tube. It should have a volume similar to the original volume of the aqueous phase (100 µl).

14. Add 50 µl of 7.5 M ammonium acetate to the tube. Precipitate nucleic acids by the addition of 320 µl of 95 percent ethanol. Mix well by inverting the tube six to ten times.

15. Place the tube in the centrifuge and orient the attached end of the lid pointing away from the center of rotation (see the icon for details). Centrifuge at maximum speed for 10 minutes at room temperature.

16. Remove the tube from the centrifuge and open the lid. Holding it by the lid, gently lift the end, touching the lip to the edge of an Erlenmeyer flask and drain the ethanol. You do not need to remove all the ethanol from the tube. **Place the tubes back into the centrifuge orienting them as before. Note:** when pouring off ethanol, do not invert the tube more than once because this can loosen the pellet.

17. Wash the pellet with 700 µl of cold 70 percent ethanol. Holding a P1000 Pipetman vertically, slowly deliver the ethanol to the side of the tube opposite the pellet. Hold the Pipetman as shown in the margin icon. **Do not start the centrifuge:** in this step the centrifuge rotor is used as a "tube holder" that keeps the tubes at an angle that is convenient for ethanol washing. Remove the tube from the centrifuge by holding it by the lid. Pour off the ethanol as described in step 16. **Note:** this procedure makes it possible to quickly wash a large number of pellets without centrifugation and vortexing. Place the tube back into the centrifuge and repeat the 70 percent ethanol wash one more time.

18. After the last wash, place the tube into the centrifuge. Make sure that the orientation is the same as before. Without closing the tube lids, start the centrifuge for 2–3 seconds to collect the remaining ethanol at the bottom of the tube. Remove all ethanol with a P200 Pipetman outfitted with a capillary tip.

19. Resuspend the DNA pellet (invisible) in 4 µl of water. This will be successful only if you know the position of the pellet on the side of the tube. It is important to realize that, for most microfuges, the pellet will be distributed on the side of the tube. To dissolve DNA, place 4 µl of water on the side wall in the middle of the tube and move the drop down to the bottom using the end of a yellow tip. Repeat this procedure several times in order to ensure that the invisible pellet at the side of the tube is dissolved. You will use all 4 µl for ligation. The final concentration of DNA solution will be 12 ng µl^{-1} (60 ng 5 µl^{-1}) to 60 ng µl^{-1} (300 ng 5 µl^{-1}) (see Table 5.1).

20. Label the tube "R- DNA" (repaired) and with your group number. Store it in a −20°C freezer. You will use this DNA for the ligation reaction in the next laboratory period.

References

Deininger, P.L. (1983) Random subcloning of sonicated DNA: application to shotgun DNA sequence analysis. *Anal. Biochem.*, **129**, 216–23.

Hengen, P.N. (1997) Method and reagents. Shearing DNA for genomic library construction. *Trends Biochem. Sci.*, **22**, 273–4.

Roe, B. and Crabtree, J.S. (1995) Protocols for recombinant DNA isolation, cloning and sequencing. In *DNA Isolation and Sequencing. Essential Techniques Series*, B.A. Roe, J.S. Crabtree and A.S. Khan (eds). John Wiley & Sons, New York.

Surzycki, S.J. (1990) A fast method to prepare random fragment sequencing libraries using a new procedure of DNA shearing by nebulization and electroporation. In *The International Conference on the Status and Future of Research on the Human Genome. Human Genome II*, p. 51. Human Genome Project, San Diego.

Surzycki, S. (2000) *Basic Techniques in Molecular Biology*. Springer-Verlag, Berlin, Heidelberg, and New York.

SECOND LABORATORY PERIOD

In this laboratory period you will carry out two of the next steps of the procedure for preparation of a sequencing library. These steps represent the procedure of DNA cloning. In the first step, you will ligate DNA fragments into the sequencing plasmid (cloning vector). In the second step, you will introduce the constructed plasmids into bacterial cells using an electroporation procedure (transformation of bacteria). A collection of bacterial cells, each carrying a single plasmid with only one DNA fragment, constitutes the sequencing library of your DNA.

Experiment 3: ligation to sequencing vector

Introduction

In vitro ligation is used for joining foreign DNA fragments to a linearized plasmid. Construction of highly representative, random fragment libraries of DNA is essential to successful sequencing using the shotgun strategy of DNA sequencing. In order to prepare these libraries it is necessary to clone repaired DNA fragments into pUC vectors, which are used as sequencing vehicles. In order to ensure that each fragment is represented in the library, the blunt-ended fragments are ligated to the *Sma*I cloning site of the pUC vectors. The *Sma*I restriction endonuclease cuts both DNA strands in the same position (CCC|GGG), thereby creating a DNA molecule with two blunt ends with a CCC sequence at the one end and a GGG sequence at the other.

Prepared vector DNA will be given to you. This vector will be restricted with *Sma*I restriction endonuclease. To prevent recirculization of the vector molecules and to select for intermolecular ligation, the vector DNA is dephosphorylated after digestion with *Sma*I restriction enzyme. Because ligase requires the presence of a 5′-phosphate group at the ends of the DNA molecules to be ligated, the dephosphorylated vector molecule ends cannot be ligated with themselves. However, repair molecules of the DNA fragment contain 5′-phosphate groups can be ligated to the vector.

Background

Principle of DNA cloning

DNA cloning can be defined in most general terms as a method of rapid isolation and amplification of DNA fragments that can be used in subsequent experiments. Cloning involves construction of hybrid DNA molecules that are able to self-replicate in a host cell, usually bacteria. This is accomplished by inserting DNA fragments into a plasmid- or bacteriophage-cloning vector, introducing the vector into bacterial cells, and amplifying vector DNA using bacterial DNA replication machinery. The DNA fragment or insert can be derived from any organism and obtained from genomic DNA, cDNA, previously cloned DNA, PCR (polymerase chain reaction) products, or synthesized *in vitro*.

Typically cloning of any DNA fragment involves the following tasks.
1. Preparing the vector for cloning.
2. Preparing DNA fragments to be cloned.
3. Joining the fragments with the vector.
4. Introducing the hybrid vector into bacteria.
5. Selecting for cells with a vector.

To accomplish these tasks, the plasmid and DNA fragment are engineered to be linear molecules with termini compatible to be joined by ligation. Then the fragment and vector are ligated together to form a circular recombinant molecule. Ligated constructs are introduced into *Escherichia coli* cells by transformation. Finally, transformed *E. coli* cells are selected from cells without a vector.

Different procedures for carrying out each of these steps have been devised. In general, these procedures can be separated into two groups: the procedures for cloning many different fragments at once, e.g. procedures for the construction of DNA libraries and the procedures for cloning one or few DNA fragments at a time. In most instances preparation of a DNA library and in particular preparation of a sequencing library requires blunt-ended cloning of DNA fragments into plasmid.

Cloning vectors

The replication machinery of bacterial cells can be used for cloning and amplifying specific DNA fragments. Most DNA fragments are incapable of self-replication in bacterial cells. However, any DNA fragment can be easily amplified and replicated when it is part of an autonomously replicating bacterial element. Most cloning vectors can be categorized according to the purpose and type of extra-chromosomal element used (Brown, 1991). The vectors used in most cloning experiments are general-purpose plasmid vectors that were made from naturally occurring procaryotic plasmids, primarily *E. coli* plasmids. These plasmid vectors have the following characteristics.

1. Cloning vectors are small, circular, double-stranded DNA molecules. The vector DNA contributes as little as possible to the overall size of recombinant molecules. This assures that a cloned fragment constitutes a large percentage of amplified and isolated plasmid DNA, making it easier to prepare large quantities of insert DNA.

2. Cloning vectors contain a **replicon**, that is a stretch of DNA that permits DNA replication of the plasmid independent of replication of the host

chromosome. This element contains the site at which DNA replication begins or the origin of replication and genes encoding RNAs and/or proteins that are necessary for plasmid replication. The replicon largely determines the copy number of the plasmid, which is defined as the number of plasmid molecules maintained per bacterial cell. High copy number plasmids are plasmids that accumulate 20 or more copies per bacteria and low copy number plasmids are plasmids that have less than 20 copies per cell (Brent and Irwin, 1989). High copy number plasmids are the most frequently used plasmids in cloning. Low copy number plasmids are useful for cloning DNA sequences deleterious to host cells or DNA sequences that are prone to rearrangement by a host, such as inverted or direct repeats. Most recombinant plasmids contain either the ColE1 or closely related pMB1 replicon (Bernard and Helinski, 1980; Brown, 1991). Examples of ColE1 replicon-containing plasmids are pBluescript or pT series plasmids and plasmids containing the pMB1 replicon are pBR322, pGEM, and pUC series plasmids. An anti-sense RNA transcript and ROP protein regulate the copy number for ColE1 and pMB1 replicons. Mutations in either of these elements can result in a higher plasmid copy number. This feature was used in the construction of very high copy number, general-purpose plasmids such as pUC and pGEM series plasmids that lack the ROP gene.

3. Cloning vectors contain selectable markers for distinguishing cells transformed with the vector from non-tranformed cells. This marker also maintains the presence of the plasmid in cells, particularly in the case of low copy number plasmids. The selectable marker is usually a gene conferring resistance to antibiotics on the host cells. This **positive selection marker** in most plasmids is the β-lactamase gene, the product of which cleaves and inactivates penicillin or its more frequently used derivative ampicillin. When transformed cells are grown in the presence of antibiotic. Cells carrying the antibiotic-resistant plasmid survive, while host cells that do not contain the plasmid are eliminated.

4. Cloning vectors contain unique cloning sites for the introduction of DNA fragments. The cloning sites in most general-purpose vectors used today consist of a **multiple cloning site** or a polylinker cloning region where a number of restriction enzyme cleavage sites are immediately adjacent to each other. These sites are chosen to be unique in the vector sequences. DNA fragments can be easily introduced into the plasmid by linearizing the plasmid by digestion with one or two enzymes present only in the polylinker region and ligating the desired fragment into it. This procedure creates a chimeric molecule without disrupting the critical features of the vector. Polylinker sites are usually flanked by sequences that can be used for priming DNA synthesis with commonly available primers for DNA sequencing or a PCR. The general purpose vectors such as pUC, pGEM, or pBluescript contain M13 reverse, M13 forward −20 and −40 primers that can be used for amplification or sequencing any inserted DNA fragment.

5. Cloning vectors contain an element for screening for the recombinant clones. The screening procedure, as opposed to selection procedures, only permits recognition of colonies transformed with vectors containing an insert from those transformed with vector alone (Rodriguez and Tait, 1983). The universally used screening method is the α-complementation screening procedure for insertional inactivation of β-galactosidase enzyme activity. The α-complementing vectors have a multiple cloning site region inserted into a DNA sequence encoding the first 146 amino acids of the *lacZ* (β-galactosidase) gene α fragment. An intact α fragment, with a few additional amino acids, can complement the ω fragment of the *lacZ* gene in bacteria restoring a fully active enzyme. The β-galactosidase enzyme metabolizes a chromogenic substrate (X-gal) and causes the formation of blue colonies on indicator plates. Inserting a DNA fragment into the polylinker region disrupts the protein-coding region of the α-fragment protein (insertional inactivation) that cannot complement into an active enzyme. Thus, cells carrying plasmid with insert have a gal$^-$ phenotype. This allows for rapid identification of bacteria containing plasmids with inserts as white colonies on indicator plates.

Preparation of plasmid for cloning

The first step in plasmid preparation is choosing the vector for cloning. The choice of vector largely depends on the experiments for which the recombinant clone will be used. For example, different types of vectors will be used for generating large quantities of DNA, expressing a fusion protein in bacteria, synthesis of mRNA, or preparation of RNA or DNA probes. Vectors used for the construction of sequencing libraries are high copy number, general-purpose vectors. A list of most of the commonly used cloning vectors is available at http://vectordb.atcg.com

The second step involves preparation of the plasmid for ligation with the insert. This step usually consists of linearization of the plasmid with an appropriate restriction enzyme and removing 5'-phosphate from the DNA ends. Because supercoiled DNA molecules are transformed at high frequencies, appropriate care must be taken to remove uncut plasmid from the preparation. Linearization of the plasmid is an enzymatic reaction and, as such, it never goes to completion. As a result, a small number of supercoiled molecules are always present and must be removed before the cloning procedure. In addition, removing the 5'-phosphate is necessary in order to limit religation of the plasmid to itself, which will create plasmid molecules lacking insert.

How plasmid and DNA fragments are prepared depends on the choice of ligation procedure to be used for joining them. There are two types of ligation procedure: ligation of fragments with compatible cohesive ends and ligation of fragments with blunt termini. Thus, there are two procedures for cloning, cohesive-end cloning and blunt-end cloning. In the preparation of sequencing libraries the blunt-end procedure is used.

Blunt-end cloning requires creating blunt termini in both the plasmid and DNA fragment since any two blunt ends are compatible for ligation. Cleaving DNA with any two restriction enzymes that create blunt ends or cleaving DNA with any restriction enzyme and then converting the protruding ends to blunt ends can generate these termini.

The use of **blunt-end cloning** has several advantages. The most obvious advantage is that any DNA fragment can be cloned regardless of its origin. Only one type of plasmid needs to be prepared for cloning all DNA fragments with blunt termini. The ligation reaction is short (10–30 minutes) and is done at room temperature. A disadvantage of the method is the necessity for preparing most of the fragments for cloning by creating blunt termini. Recent improvements in blunt-end cloning technology have resulted in increases in blunt-end cloning efficiency to approximately 10^7–10^9 transformants per microgram of DNA, removing the most important disadvantage of this method.

During ligation, DNA ligase will catalyze the formation of phosphodiester bonds between adjacent nucleotides only if phosphate is present at the 5′-end of DNA molecules. Removing the 5′-phosphates from both ends of the linear plasmid DNA and leaving them on the fragment to be cloned will minimize recircularization of the plasmid. The enzyme used for this purpose is either bacterial alkaline phosphatase or calf intestinal phosphatase. The result of this treatment is that neither strand of the vector can form a phosphodiester bond if insert DNA is not present in the reaction mixture.

Ligation reaction

Ligation or joining of a foreign DNA fragment to a linearized plasmid involves formation of new bonds between the 5′-phosphate and 3′-hydroxyl ends of DNA. There are two enzymes available for catalysis of this reaction *in vitro*: *E. coli* DNA ligase and bacteriophage T4 DNA ligase. For almost all cloning purposes the bacteriophage enzyme is used. Ligation of dephosphorylated plasmid ends with a phosphorylated insert results in the formation of hybrid molecules with two phosphate bonds and two single-stranded gaps. These gaps are repaired after introduction of plasmid into bacterial cells.

The most significant factor in a ligation reaction is the concentration of DNA ends. Essentially, during ligation there are two competing reactions: bimolecular concatamerization and unimolecular cyclization. In order to achieve successful cloning, the first reaction to occur should be bimolecular concatamerization of the vector with insert followed by a recirculization reaction to form circular plasmid. Unfortunately these two reactions are mutually exclusive, so some form of compromise must be used.

The ratio of recirculization and concatameric ligation products is dependent on two factors i and j, where i is the total concentration of DNA termini in the reaction mixture and j is the effective concentration of one end of a DNA molecule in the immediate neighborhood of the other end of the same molecule. The value of j is constant for a linear DNA molecule of a given length and is **independent of DNA concentration**. For cohesive-end ligation and blunt-ended ligation, the value of i in ends per milliliter can be calculated from the equation

$$i = 2N_0 M \times 10^{-3} \text{ ends ml}^{-1} \tag{5.2}$$

where N_0 is the Avogadro number and M the molar concentration of the DNA molecules.

The j value can be determined using the equation (Dugaiczyk et al., 1975)

$$j = j_\lambda \left(\frac{mw_\lambda}{mw_X} \right)^{3/2} \text{ ends ml}^{-1} \tag{5.3}$$

where, j_λ is 3.6×10^{11}, mw_λ is the molecular weight of the λ genome (30.8×10^6 Da), and mw_x is the molecular weight of unknown molecule. After converting M to DNA concentration in micrograms per milliliter [DNA] and mw_x to kbp, the ratio $j:i$ for any given DNA is equal to (rearranged from Rodriguez and Tait, 1983)

$$j/i \approx \frac{51.1}{0.812 (kbp_x)^{0.5} [DNA]} \tag{5.4}$$

Equation (5.4) indicates that, to achieve a low $j:i$ ratio, the concentration of DNA in the reaction should be high. For any ligation reaction, three conditions apply.
1. When $i = j$ there is an equal chance of one end of the molecule making contact with the end of a different molecule and/or the end of the same molecule.
2. When $j > i$ there is a greater chance of one end of the molecule contacting the other end of the same molecule than of finding the end of another molecule. Therefore, under this condition the recirculization reaction predominates.
3. When $j < i$ there is a greater chance of one end of the molecule contacting the other DNA molecule rather then its own end. Therefore, under this condition, the concatamerization reaction predominates.

Although the theoretical consideration discussed above predicts that conversion from concatamerization to circulization should occur at $j:i = 1$, experimental observation indicates that this conversion actually occurs at $j:i = 2-3$ (Dugaiczyk et al., 1975).

To choose the correct compromise between two types of ligation one should realize that a ligation reaction is not an instantaneous event. This reaction proceeds through a progressive series of individual ligation events, where each event significantly alters the $j:i$ ratio for the remaining unligated molecules. Since the value of j is constant, as the reaction proceeds the value of i decreases resulting in increases of the $j:i$ ratio. The high $j:i$ ratio favors the formation of circular molecules rather than linear concatamers.

For cloning DNA insert into plasmid, the reaction conditions initially should favor joining the insert to one end of the plasmid vector. The second ligation event, occurring later, should be circularization of the plasmid–insert hybrid molecule to avoid joining the next linear fragment to the first one. To achieve this, the ratio of $j:\Sigma i$ at the beginning of ligation should be low to allow the concatamerization reaction and, as the reaction proceeds, this ratio should rise to a value greater than three to permit circularization of the plasmid. This can be done in three ways.

1. Starting the ligation reaction at a high concentration of vector DNA. As equation (5.4) indicates, a low $j:i$ value depends on the DNA concentration per milliliter. For the vectors commonly used for cloning (2,600–4,000 bp), the concentration of vector DNA is usually adjusted in the range 3–6 µg ml^{-1} or 1–3 pmol ml^{-1}.

2. Adjusting the initial molar ratio of plasmid to linear fragments to be greater than 1.0. Since during the reaction the j value is not changed, but the i value will decrease as a result of the ligation of one end of the fragment to the plasmid, the $j:i$ ratio will increase towards the end of the reaction. At a high $j:i$ ratio the circularization reaction predominates, resulting in the formation of circular hybrid molecules. The insert concentration is usually adjusted to a two- to fourfold **molar** excess over plasmid DNA. The $j:i$ ratio for such a reaction can be approximately calculated using equation (5.4) and assuming a kbp average between vector and insert. For example, using pUC19 plasmid in the ligation reaction (2.68 kbp) at concentration 6 µg ml^{-1} and 2.6 times molar excess of 1 kbp insert (6 µg ml^{-1}), the $j:i$ ratio at the beginning of the reaction is approximately 4.0 (51.1/(2.68 + 1.0/2 bp)$^{0.5}$ × 12 µg ml^{-1} × 0.812 = 3.8). This ratio will steadily increase as the reaction proceeds resulting in almost complete circulization of the hybrid plasmid molecules.

3. Using phosphatase-treated vector with a phosphorylated insert. Under this condition the vector cannot self-ligate at any $j:i$ ratio and, therefore, the initial concentration of plasmid is not important. The initial concentration of the insert is chosen so as not to allow its circularization at the beginning of the reaction. Ligation of fragment to vector results in the formation of a new molecule that can be self-ligated into a circle. Cyclization of a hybrid molecule depends on its j value and is independent of insert concentration as long as the molar concentration of the insert is equal to or higher than the molar concentration of plasmid. For all useful molar ratios of vector to insert (1:1 to 1:3) the $j:i$ for insert is always high, favoring circularization of the insert rather than concatamerization. Thus, cloning of concatamers of the insert does not occur. Using an insert concentration higher than 1:3 is not recommended because it may lead to cloning inserts that have been ligated together. Following formation of the hybrid molecules, conditions favoring cyclization of the hybrid predominate if the insert is not larger than the vector (a high $j:i$ ratio for hybrid). Practically, lower concentrations of plasmid are preferred because this leads to nearly complete formation of circular hybrid molecules. For example, ligation of plasmid at 2.5 µg ml^{-1} with a 2.68 kbp insert gives a $j:i$ ratio for the construct equal to 5.4, whereas the same ligation at 1 µg ml^{-1} plasmid gives value $j:i$ equal to 13.6.

The efficiency of a ligation reaction depends not only on DNA and enzyme concentrations, but also on the purity of the vector and insert, the reaction pH and temperature, and the presence of inhibitors of ligase. In general, most deleterious components that frequently contaminate ligation reactions are as follows.

1. Monovalent salts (for example sodium chloride) or ammonium salts. Concentrations of sodium chloride above 50 mM and ammonium salt above 10 mM severely inhibit ligation reactions.

2. Phosphate. Blunt-end ligation is particularly sensitive to this salt and concentrations of phosphate greater than 25 mM should be avoided.

3. ATP concentration. Blunt-end ligation is inhibited by ATP concentrations above 0.1 mM.

Technical tips

Ligation reactions can fail for several reasons (listed in frequency of occurrence).

1. Too high a concentration of vector. This can happen because the concentration of fragments can be lower than expected due to low efficiency of the nebulization process. It is possible to repeat the ligation reaction using a lower amount of vector (2–5 ng per reaction) and the same amount of DNA fragments.

2. The presence of a high concentration of monovalent salts (for example sodium chloride or ammonium salts). The concentrations of these salts in the final reaction should not exceed 50 mM. Careful washing of DNA pellets with 70 percent ethanol should remove all salts from plasmid and insert preparations. Neglecting to do so is a frequent cause of failure in ligation reactions.

3. Too high a concentration of ATP. The blunt-end ligation reaction is inhibited by the presence of an ATP concentration above 0.1 mM. Buffers supplied by manufactures with T4 DNA ligase frequently contain ATP at optimal concentration for cohesive end ligation. The use of these buffers in

Table 5.2 Master reaction mixture

Ingredients	Add to reaction	Final concentration
Two times ligation buffer (vial 1)	50.0 μl	One times
Five times DNA buffer (vial 2)	20.0 μl	One times
pUC18 DNA (10 ng/μl)	10.0 μl	20 ng (10 fmol)

Table 5.3 Ligation reactions

Ingredients	C1	C2	3	4
Master reaction mixture	16.0 μl	16.0 μl	16.0 μl	16.0 μl
Your DNA (tube R)	–	–	4.0 μl	–
Partner's DNA (tube R)	–	–	–	4.0 μl
Water	5.0 μl	4.0 μl	–	–
T4 DNA ligase (vial 3)	–	1.0 μl	1.0 μl	1.0 μl

blunt-end ligation reactions is a very frequent error, causing poor ligation efficiency. Carefully check the content of the ligation buffer supplied when performing blunt-end ligation.

4. Degradation of ATP in the stock solution or old ligation buffer. Always use freshly prepared 0.5 mM ATP.

Protocol

You will perform two ligation reactions using the repaired DNA from both you and your partner prepared during the first laboratory period. The protocol includes two control reactions. The first control reaction tests for the presence of undigested vector DNA in plasmid preparations. The second control tests the efficiency of dephosphorylation of vector ends.

1. Label one 1.5 ml microfuge tube RX. Also prepare two control reaction tubes labeled C1 and C2 and two ligation reaction tubes labeled 3 and 4. Place the tubes on ice.
2. Prepare the master reaction mixtures for five reactions in the RX tube as shown in Table 5.2.
3. Mix ingredients by pipetting up and down. Centrifuge for 5 seconds in a microfuge. Place tubes on ice. **Note:** it is absolutely necessary to mix the contents of vials 1 and 2 thoroughly directly before use.
4. Place tubes C1, C2, 3, and 4 at room temperature and prepare reactions as indicated in Table 5.3. Add water, the master reaction mixture, and DNA first. Mix the ingredients gently by pipetting up and down several times. Centrifuge for 5–10 seconds to collect liquid at the bottom of the tubes.
5. Start the reaction by the addition of enzyme to tubes C2, 3, and 4. Mix

well by gently pipetting up and down several times. Incubate the reactions for 10 minutes at room temperature. **Note:** the expected molar ratio between vector (1.6 kb) and insert (1–2 kb) is approximately 3 : 1 to 15 : 1 assuming all repaired DNA was recovered.

6. Stop the reactions with the addition of 1 µl of 0.5 M EDTA, pH 8.0. Add 80 µl of water to the reaction tubes and mix well by inverting the tubes several times.

7. Add 50 µl of 7.5 M ammonium acetate to each tube and mix well by inverting several times.

8. Add 300 µl of 95 percent ethanol to each tube and mix well by inverting four to five times. Place the tubes in the microfuge and centrifuge for 20 minutes at room temperature. **Be sure to orient each tube in the centrifuge rotor with the lid closing pointing away from the center of rotation.** This will "mark" the position of pelleted DNA since your pellet will not be visible.

9. Remove the tube from the centrifuge and open the lid. Gently lift the end, touching the tube to the edge of an Erlenmeyer flask and drain the ethanol. You do not need to remove all the ethanol from the tube. Place the tubes back into the centrifuge in the same orientation as above. **Note:** when pouring off ethanol do not invert the tube more than once because this can loosen the pellet.

10. Wash the pellet with cold 70 percent ethanol. Add 700 µl of 70 percent ethanol to the tube, using a P1000 Pipetman. Holding the P1000 Pipetman vertically (see the icon in the margin) slowly deliver the ethanol to the side of the tube opposite the pellet. **Do not start the centrifuge.** In this step the centrifuge rotor is used as a "tube holder" that keeps the tube at an angle convenient for ethanol washing. Withdraw the tube from the centrifuge by holding the tube by the lid. Remove ethanol as in step 9. **Note:** this procedure makes it possible to quickly wash a large number of pellets without centrifugation and vortexing. Vortexing and centrifuging the pellet are time-consuming and frequently lead to substantial loss of material.

11. Place the tube back into the centrifuge and repeat the 70 percent ethanol wash one more time.

12. After the last wash, place the tube into the centrifuge, making sure that the tube position in the rotor is the same as in steps 9 and 10. Without closing the tube lids, start the centrifuge for 2–3 seconds and collect the remaining ethanol at the bottom of the tube. Remove all ethanol with a P200 Pipetman outfitted with a capillary tip. **Note:** never **dry the DNA pellet** in a vacuum. This will make dissolving the DNA pellet very difficult if not impossible.

13. Resuspend the DNA pellet (invisible) in 5 µl of water. This will be successful only if you know the position of the pellet on the side of the tube. It is important to realize that, for most microfuges, the pellet will be distributed on the side of the tube. To dissolve DNA, place 5 µl of the water on the side wall in the middle of the tube and move the drop down to the bottom using the end of a yellow tip. Repeat this procedure several times to assure that the invisible pellet at the side of the tube is dissolved.

References

Bernard, H.U. and Helinski, D.R. (1980) Bacterial plasmid cloning vectors. In *Genetic Engineering. Principles and Methods*, Vol. 2, J.K. Setlow and A. Hollaender (eds), pp. 133–67. Plenum Press, New York and London.

Brown, T.A. (1991) Cloning vectors. In *Molecular Biology Labfax*, B.D. Hames and D. Rickwood (eds), pp. 193–234. Bios Scientific Publishers Ltd and Academic Press Inc., Oxford.

Dugaiczyk, A., Boyer, H.W., and Goodman, H.M. (1975) Ligation of *Eco*RI endonuclease-generated DNA fragments into linear and circular structures. *J. Mol. Biol.*, **96**, 171–84.

Brent, R. and Irwin, N. (1989) Introduction to plasmid biology. In *Current Protocols of Molecular Biology*, F.M. Asubel, R. Brent, R.E. Kingston, D.D. Moore, J.G. Seidman, J.A. Sith et al. (eds), pp. 1.5.1–1.5.8. John Wiley & Sons, Inc.

Rodriguez, R.L. and Tait, R.C. (1983) *Recombinant DNA Techniques: An Introduction*. Addison-Wesley Publishing Co., Reading, MA.

Experiment 4: transformation of bacteria by electroporation

Introduction

In this experiment you will prepare a sequencing library by transforming *E. coli* cells with plasmids containing random fragments of your DNA. You will use the entire product of the ligation reactions prepared in experiment 3. Each group will carry out four transformations using two control ligation products and two experimental ligation products. You will use 30 µl of cells for each transformation by electroporation (cell concentration 10^{10} cell ml^{-1}) and an electroporation cuvette with a 0.1 cm electrode gap. Using this method the frequency of transformation is 0.02. Low transformation frequency prevents co-transformation of a cell with two or more plasmid molecules.

The bacterial cells used in this experiment will be given to you ready for transformation. You will plate transformed cells on a selective medium that permits growth of transformed colonies and visual selection of cells with plasmid containing insert (ampicillin and X-gal IPTG). The colonies will be used to prepare plasmid for DNA sequencing.

The electroporation method of transformation has numerous advantages over other transformation methods and is always used for preparation of sequencing libraries. The advantages of this method are as follows.

1. Electroporation yields very high transformation frequencies that come close to the frequencies obtained when using bacteriophage vectors. At saturating DNA concentrations, 80–100 percent of the cells are transformed.
2. The technique makes it possible to transform very small volumes of cells. A volume of cells as small as 20 μl can be transformed yielding approximately 10^9 transformants.
3. Preparation of cells for transformation is very simple and does not use elaborate and time-consuming protocols. Moreover, the cells used for electroporation can be prepared ahead of the time and stored indefinitely without losing competence.
4. Transformation efficiency is practically independent of DNA size and form. Linear DNA molecules, supercoiled DNA, and circular DNA are transformed with equal efficiency.

Background

Electroporation is the most efficient transformation procedure that can be used for bacterial transformation. It involves a brief application of high-voltage electric field to the cells resulting in the formation of transient holes in the cell membrane through which plasmid DNA can enter the cell. The method was originally developed for animal and plant cells (Neumann et al., 1982) and later for bacteria (Böttger, 1988; Dower et al., 1988; Li et al., 1988; Smith et al., 1990). Transformation efficiencies as high as 10^{10} transformants μg^{-1} of plasmid have been achieved for *E. coli* cells (Dower et al., 1988; Smith et al., 1990).

The maximum efficiency of this process depends on many variables. The most important are electrical pulse shape, electrical field strength, and electrical pulse time. The relationship between these parameters is described by the equations (Shigekawa and Dower, 1988)

$$V_t = V_0 [e^{-(t/\tau)}] \tag{5.5}$$

$$E = \frac{V_0}{d} \tag{5.6}$$

$$\tau = R \times C \tag{5.7}$$

where V_t is voltage at time t, V_0 is the initial voltage or peak of voltage, τ is the pulse time constant, R is the resistance of the circuit (ohms), C is the capacitance of the charged capacitor (farads), E is the electric field strength, and d is the distance between the electrodes.

In most instruments used today, charging a capacitor to a predetermined voltage and its subsequent discharge to the electroporation chamber generates an electric pulse. Capacitor discharge results in an exponentially decaying pulse shape that is characterized by the value of τ. According to equation (5.5), τ can be defined as the time over which the voltage declines to $1/e$ or approximately 37 percent of the initial value (V_0). The electrical field is derived from the peak voltage (V_0) delivered to the chamber at the moment of discharge. At this moment the difference in potential in the chamber is at its maximum, generating membrane depolarization and pore formation. The formation of pores is largely dependent on the field strength parameter determined by equation (5.6). The introduction of external compounds (e.g. plasmids, DNA, proteins, etc.) into the cell critically depends on the voltage drop determined by τ.

The time constant τ is specific for each cell type and, in general, the smaller the cell size, the shorter the τ needed to introduce external elements. The optimal time constant for *E. coli* was determined to be 5 milliseconds. The field strength for optimal electroporation of different bacterial strains differs and must be experimentally determined.

Transformation results are usually described by the values representing the frequency of transformation (f_{tr}) and efficiency of transformation (E_{tr}). Equation (5.8) describes the frequency of transformation and the efficiency of transformation is described by equation (5.9).

$$f_{tr} = \frac{Tr}{S} \tag{5.8}$$

$$E_{tr} = \frac{Tr}{[pDNA]} \tag{5.9}$$

where *Tr* is the number of transformants, *S* is the number of cells at 80 percent survival, and [pDNA] is the concentration of plasmid DNA used in micrograms. The frequency of transformation is directly dependent on DNA concentration in the range of $10\,pg\,ml^{-1}$ to $7.5\,\mu g\,ml^{-1}$ and, at these DNA concentrations, 80 percent of the cells survive electroporation (Dower et al., 1988). Moreover, because from equation (5.8), $Tr = f_{tr} \times S$, the efficiency of transformation can be rewritten as

$$E_{tr} = \frac{f_{tr} \times S}{[pDNA]} \tag{5.10}$$

Equation (5.10) indicates that the transformation efficiency (E_{tr}) is high when the cell concentration is high (10^9 to 3×10^{10}) and the DNA concentration is low ($1-10\,ng\,rx^{-1}$). These conditions are commonly used in electroporation of *E. coli* in order to achieve high transformation efficiencies and avoid co-transformation. For example, for a standard electroporation reaction, 30 μl of cells at a concentration of 10^{10} cells ml^{-1} are used with 5 ng of plasmid DNA. The transformation efficiency commonly achieved is 10^9 transformants μg^{-1} of DNA, giving a transformation frequency of two transformants per 100 surviving cells ($f_{tr} = E_{tr} \times$ [pDNA]/S) ($f_{tr} = 10^9 \times 0.005/2.4\,10^8 = 0.02$). At this transformation frequency, transformation of a single cell with two or more plasmids is very rare.

In addition to the parameters described above, electroporation is affected by a number of other variables such as temperature, the components of the electroporation medium, and the method of cell recovery after electroporation.

Electroporation at low temperature is approximately 100 times more efficient than when cells are pulsed at room temperature. Low temperature presumably affects the fluidity of the membrane, aiding in pore formation and slowing their closure. Similarly quick restoration of membrane fluidity and closing pores is crucial for cell survival after the pulse. Thus, cells should be transferred to prewarmed growth medium as fast as possible after pulse application. Delaying this transfer by 1 minute will cause a threefold drop in the transformation efficiency and cell survivability. Moreover, cells should be given some time to recover and rebuild their membrane before plating on solid medium. The recovery time should be short enough to prevent cell division that will result in the "cloning" of transformed cells. This is particularly important when electroporation is used for the preparation of genomic and sequencing libraries. A recovery time of 45–50 minutes at 37°C appears to be the best for cell recovery and short enough to not allow any cell division.

Technical tips

In this experiment we will use a special electrocompetent strain of *E. coli*, ElectroMax DH10B. Many other electrocompetent strains can be used, as long as they support α-complementation, for distinguishing bacterial cells containing plasmid with insert from those without it. The bacterial strain should also be ampicillin sensitive.

Low or no transformation results if arcing occurs during pulsing. Arcing can occur for the following reasons.
1. There is residual salt or buffer in the sample due to inadequate washing of ligation reactions with 70 percent ethanol.
2. There is the presence of air bubbles in the sample due to incorrect pipetting into the electrophoresis chamber.
3. There is too high a concentration of cells used in electroporation. This would result in arcing of all samples. Electroporate cells without plasmid are added to test this possibility.
4. There are old cell preparations, incorrectly stored cells, or the cells are thawed too fast. All of these can bring about partial lysis of cells causing arcing.

5. There is too small a volume of cells in the electroporation chamber. A cuvette with a 0.1 cm electrode gap requires at least 25 µl of sample.

A lower than expected frequency of transformation can result from the following.

1. Incorrect setting of the electroporation apparatus with too high or too low a voltage than that recommended for a given bacterial strain. The voltage recommended with ElectroMax DH10B is 1850 V.
2. An incorrect τ constant. This value can be adjusted on some electroporation units. Choose the correct capacitance of the capacitor (in farads) and resistor (in ohms) to give a τ constant close to 5 milliseconds. Use equation (5.7) for calculating this value. For example, if using an Invitrogen Electroporator II choose a 50 µF capacitor and 150 Ω resistor in order to obtain a τ constant of 7.5 ms (50×10^{-6} F $\times 150 \Omega = 0.0075$ s).
3. A warm electroporation chamber or sample during pulsing. Warming frequently occurs during loading of the samples. It is better to chill the electroporation cuvettes after loading the sample even if this prolongs incubation of the cells with DNA over the recommended time.
4. Excessive volume of the cells in the electroporation chamber. Cell volume should not exceed 40 µl.

Another problem frequently met is the presence of all blue colonies on experimental plates 3 or 4 with concomitant absence of blue colonies on both control plates (C1 and C2). This can happen when the insert is cloned in a frame with the α-peptide or when the insert DNA is small (<200 bp). In the last instance, color development will be weak and the colonies may appear pale blue or white. Colonies with recombinant plasmid can be sometimes distinguished from those with plasmid alone by colony morphology. Recombinant colonies appear translucent while non-recombinant colonies are opaque. The cells from such colonies can be used for plasmid isolation and sequencing.

Small colonies surrounding a single colony (satellite colonies) will frequently appear on ampicillin plates. These colonies appear near ampicillin-resistant colonies because the β-lactamase enzyme, which is responsible for antibiotic resistance, is secreted from the cell, thereby removing antibiotic from the agar in the vicinity. This usually results from two low a concentration of antibiotic in the plate or the plate being incorrectly stored. Ampicillin-containing plates should be stored at 4°C for no longer than three to four weeks. The satellite colonies can be eliminated by short incubation of the plates (less than 14 hours). In this case in order to enhance blue color development without further bacteria growth the plates can be incubated at 4°C for several hours. Because the cells for plasmid isolation will be grown in the medium supplemented with ampicillin, satellite cells will not grow even if accidentally transferred together with transformed cells.

Electroporation cuvettes can be reused three to five times. The efficiency of transformation progressively decreases with successive use of a cuvette. This is probably due to some cells being "baked" onto the electrode surface or pitting of the electrode surface by washing procedures. Electroporation cuvettes can be washed before reusing using the following procedure.

1. Immediately after use immerse the cuvette in 1 percent Alconox solution. This will prevent drying of bacterial cells onto the electrode surface.
2. Rinse the cuvette six to eight times with distilled or deionized water. Do not keep cuvettes in Alconox solution for more than 1 hour.
3. Rinse the cuvette three times with 70 percent ethanol. Fill the cuvette with ethanol, cap it, and invert it several times. This treatment should sterilize the cuvette.
4. Dry the cuvette by filling it with 95 percent ethanol, inverting it several times, pouring off the ethanol, and drying it upside-down on a paper towel.

"Falcon 2059" polypropylene culture tubes should be used for cell recovery after electroporation. The use of glass tubes will result in poor recovery. It is possible to use other brands of polypropylene tubes, but some batches can occasionally be contaminated with surfactants that inhibit transformation. The manufacturer tests Falcon 2059 tubes for such contamination.

Protocol

Before starting the electroporation procedure, make the following preparations: (i) label four electroporation cuvettes C1, C2, 3, and 4 and cool them on ice for at least 5 minutes; (ii) label four Falcon 2059 tubes with the numbers corresponding to those of the electroporation cuvettes and place these tubes in the tube rack at room temperature; (iii) thaw the ligation reaction (3 and 4) and two control reactions (C1 and C2) prepared previously; (iv) set the electroporator voltage to 1.85 kV; and (v) warm up 10–50 ml of TB medium to 37°C.

1. Remove the bacterial cells from −70°C storage and gently thaw them on ice. Tap the tube gently to mix the cells. **Note:** do not leave cells on ice for an extended time. Use the cells as soon as possible. Cells can be refrozen for later use, but the transformation frequency will be significantly lower.
2. Add 30 µl of cells to the tube labeled C1 (control). Pipette the cells up and down two times to mix the DNA with the cells. Be very careful not to create air bubbles during this procedure.
3. Transfer 30 µl of this mixture to a cuvette labeled C1. Holding the cuvette at a 45° angle, deliver the mixture to the lower end of the electroporation chamber of the cuvette. Deliver liquid slowly and do not operate the Pipetman beyond the first stop. This procedure will prevent the formation of air bubbles in the electroporation chamber. Tap the cuvette on the laboratory bench several times to distribute the liquid on the bottom of the chamber. Close the cuvettes with the cap provided and place it back on

ice for 45 seconds. **Note:** an electroporation chamber with a 0.1 cm gap is very narrow. Frequently the cell sample will stay at the top of the chamber. Vigorous tapping of the cuvette on the laboratory bench will make sample flow to the bottom. It is important to do this as quickly as possible and not warm up the cuvette and cells.

4. Place the cuvette into the electroporation machine and initiate electroporation. **Note:** if a loud "snap" is heard while pulsing, arcing occurred inside the electroporation chamber. Continue with the protocol as usual; some transformation may still have occurred. See the section on technical tips for how to prevent arcing.

5. As quickly as possible add 1 ml of warm TB medium directly into the cuvette using a P1000 Pipetman. Pipette slowly up and down two times. Gently transfer all the cells from the electroporation cuvette into the appropriately labeled Falcon tube. **Treat the cells very gently: they are very fragile after electroporation.** Leave the tube at **room temperature** until all samples have been electroporated. **Note:** do not discard electroporation cuvettes. They can be reused after washing for less critical transformations (for example control transformations). A protocol for washing electroporation cuvettes is given in the section on technical tips.

6. Electroporate samples C2, 3, and 4 following the procedure described in steps 2–5.

7. Transfer all Falcon tubes to an orbital shaker and incubate the cells at 37°C for 45 minutes in order to allow the cells to recover and to express antibiotic resistance. The rotating speed of the shaker should not exceed 240 r.p.m.

8. Plate two plates, one with undiluted cells (1:1) and one with diluted cells (1:10), respectively. To prepare the 1:10 dilution add 900 μl of PBS to four microfuge tubes and label each tube with the sample name and dilution factor (for example 3 1:10, 4 1:10, C1 1:10, etc.). Prepare two plates for each electroporation reaction labeled on the bottom of the plate with the sample name and appropriate dilution (for example C1 1:1, C1 1:10, 3 1:1, etc.). **Note:** for optimal cell survival, use only PBS for dilution. Diluting the cells in growth medium instead of PBS will lower cell viability by approximately 20 percent.

9. Mix the cells in the tube labeled C1 by gently tapping with fingers. Transfer 100 μl to the dilution tube labeled C1 1:10. This will constitute 1:10 dilution of the original cell culture. Mix well by pipetting up and down several times.

10. Using the same yellow tip, immediately pipette 100 μl of cells from the 1:10 dilution onto the corresponding plates.

11. Using the same tip, transfer 100 μl of cells from Falcon tube C1 onto the plate labeled C1 1:1. **Note:** make one set of dilutions at a time. Finish plating the bacteria from a single electroporation tube before preparing the next dilution set. This prevents cross-contamination and plating errors.

12. Sterilize the cell spreader by dipping it into a beaker of 95 percent ethanol and briefly pass it through a flame to ignite the alcohol. Burn off the ethanol keeping the spreader **away** from the burner flame.
13. Cool the spreader by touching it to the agar away from the cells. Spread the cells on a plate with a 1:10 dilution first by dragging the cell suspension across the agar surface with the spreader back and forth several times. Spread the cells on a plate labeled 1:1 next. Return the spreader to the beaker with ethanol.
14. Replace the plate lids and let them stand until all liquid is absorbed into the agar.
15. Dilute and plate the cells from the remaining electroporation tubes using the procedure described in steps 9–13. Plate the cells from the 3 and 4 tubes last.
16. Place the plates upside down in a 37°C incubator and incubate them for 15–18 hours.

Next day

1. Count the cells on plates 3 and 4. Count only white colonies. Calculate the efficiency of transformation, which is expressed as the number of transformants per microgram of plasmid used. Use the following equation for this calculation: E_{tr} = number of transformed colonies/nanograms of plasmid DNA in the ligation rx × dilution factor.
2. Inspect two control plates C1 and C2. They should contain a few blue colonies. The presence of a large number of colonies on both plates indicates a failure to linearize the plasmid during vector preparation and/or incomplete dephosphorylation of its 5′-ends. If this is the case, the plate with the ligated insert and plasmid (3 and 4) will also contain mostly blue colonies.
3. Prepare five bacterial cell cultures for plasmid preparation. Each student should prepare five cultures. You will be given 10 ml glass tubes containing 2 ml of TB-amp medium (100 µg ml^{-1}). Inoculate each tube with a single white colony from plates 3 and 4. Using a sterile toothpick, touch the colony and drop the toothpick into the tube. Grow cells overnight on an orbital shaker and store cells at 4°C until the next laboratory period.

Expected results

Figure 5.4 shows a plate with a 1:10 dilution of a human sequencing library prepared using the protocol described. The expected efficiency of transformation is approximately 10^6–10^7 transformants per microgram of plasmid used. This will result in approximately 100–1,000 colonies on a 1:10 dilution plate. If the transformation efficiency is very low there should be at least ten to 100 colonies on a plate with a 1:1 dilution.

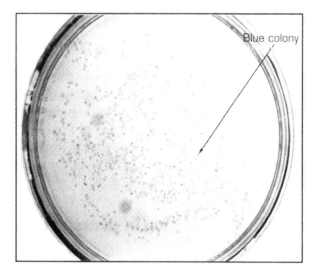

Figure 5.4 Human sequencing library plated at 1 : 10 dilution. After electroporation, cells were grown for 45 minutes in 1 ml of TB medium. Before plating the cells were diluted 1 : 10 in one times PBS and 100 µl were plated onto LB agar medium containing X-gal and IPTG and 100 µg ml^{-1} ampicillin. The plates were incubated overnight at 37°C.

References

Böttger, E.C. (1988) High-efficiency generation of plasmid cDNA libraries using electrotransformation. *BioTechniques*, **6**, 878–80.

Dower, W.J., Miller, J.F., and Ragsdale, C.W. (1988) High efficiency transformation of *E. coli* by high voltage electroporation. *Nucleic Acids Res.*, **16**, 6127–45.

Li, S.J., Landers, T.A., and Smith, M.D. (1988) Electroporation of plasmids into plasmid-containing *Escherichia coli*. *BioTechniques*, **12**, 72–4.

Neumannn, E., Schaefer-Ridder, M., Wang, Y., and Hofschneider, P.H. (1982) Gene transfer into mouse L-cells by electroporation in high electric field. *EMBO J.*, **1**, 841–5.

Shigekawa, K. and Dower, W.J. (1988) Electroporation of eukaryotes and prokaryotes: a general approach to the introduction of macromolecules into cells. *BioTechniques*, **6**, 742–51.

Smith, M., Jesse, J., Landers, T.A., and Jordan, J. (1990) High efficiency bacterial electroporation: 1×10^{10} *E. coli* transformants/µg. *Focus*, **12**, 38–40.

THIRD LABORATORY PERIOD

In this laboratory period you will isolate plasmid using modification of alkaline procedure (Surzycki, 2000) and run sequencing reactions. In the second laboratory period you prepared a sequencing library and grow bacterial cells from a single bacterial colony. This colony started from as a single transformed cell in your sequencing library. Consequently, all of the cells contain an identical plasmid. This plasmid, when purified, can be used for sequencing the cloned insert.

Experiment 5: preparation of plasmid for DNA sequencing

Introduction

In this experiment you will isolate plasmids that will be used for DNA sequencing. The plasmids will be purified from liquid cultures prepared in the second laboratory period. These cells contain plasmids from your genomic DNA sequencing library. The method used will be a modification of the alkaline lysis procedure that is suitable for the fast preparation of plasmid template for sequencing. The modification encompasses neutralization of the alkaline lysate with ammonium acetate rather than potassium or sodium salts and removal of the remaining proteins by precipitation with 1.87 M ammonium acetate. This allows purification of plasmid DNA without the use of toxic organic solvents. The procedure is quick and yields plasmid DNA of a purity comparable to the CsCl-prepared DNA.

Background

Isolation of plasmid DNA from bacterial cells is an essential step for many molecular biology procedures. Many protocols for large- and small-scale isolation of plasmids (mini-preps) have been published. The plasmid purification procedures, unlike the procedures for purification of genomic DNA, should involve removal not only of protein, but also another major impurity: bacterial chromosomal DNA. The task of plasmid purification differs substantially from that of the preparation of genomic DNA. Most plasmid DNA purification methods start from the preparation of a crude bacterial lysate and eventually employ standard protein removal procedures. To achieve separation of plasmid from chromosomal DNA, these methods exploit the structural differences between plasmid and chromosomal DNA. Plasmids are circular supercoiled DNA molecules substantially smaller than bacterial chromosomal DNA.

There are three basic methods of plasmid preparation.

1. Alkaline lysis, which was introduced by Birnboim and Dolly (1979) and Birnboim (1983). In this method cells are lysed and DNA denatured by sodium deodecyl sulfate (SDS) and NaOH. Neutralization of the solution results in a fast reannealing of covalently closed plasmid DNA due to the interconnection of both single-stranded DNA circles. Much more complex bacterial chromosomal DNA cannot reanneal in this short time and forms a large insoluble DNA network largely due to interstrand reassociation at multiple sites along the long linear molecules. In the next step of the procedure, lowering the temperature results in the precipitation of protein–SDS complexes. Subsequently, both complexes, DNA and protein, are removed by centrifugation leaving plasmid molecules in the supernatant. If cleaner plasmid is desired, the remaining protein and RNA are removed by standard methods.
2. Lysis by boiling in the presence of detergent, which was introduced by Holmes and Quigley (1981). In this method high temperature and detergent lyse bacteria cells. Bacterial chromosomal DNA under these conditions remains attached to the bacterial membrane. Subsequent centrifugation pellets chromosomal–DNA complexes while plasmid DNA remains in the supernatant. A recent modification of this procedure involves the lysis of bacterial cells using a microwave oven rather than a boiling water bath (Hultner and Cleaver, 1994; Wang et al., 1995). Further plasmid purification, if desired, can be carried out using standard deproteinization procedures and RNase treatment.
3. Application of affinity matrixes for plasmid or proteins, which was introduced by Vogelstein and Gillespie (1979).

Technical tips

The success of plasmid isolation using alkaline lysis depends on three critical steps.
1. Resuspension of cells in solution II. Solution II should be added to all tubes at once at room temperature. The cells should be resuspended one tube at the time and placed on ice. The time of incubation on ice is not critical and the stipulated time is the minimal time necessary. Cells should be resuspended in solution II thoroughly, resulting in a uniform, very viscous solution.
2. The addition of ammonium acetate solution. This solution should be as fresh as possible in order to assure complete neutralization of the lysate. Incomplete neutralization will result in contamination with bacterial chromosomal DNA. This will make sequencing of the plasmid impossible.
3. Cells should be grown in TB medium overnight, i.e. 17–20 hours. Prolonged growth of the cells on TB medium can result in lysis of the cells. If a medium other than TB is used for growth (e.g. L-broth), the volume of cells used for preparation should be increased to 500–600 μl.

Some plasmid preparations can be contaminated with RNA. This contamination appears as a diffuse band running together or slightly ahead of the bromophenol blue dye during gel electrophoresis of plasmids. For most sequencing protocols, some contamination with RNA is not deleterious. However, it is better to sequence a plasmid preparation with little or no RNA.

It is important to follow the ethanol washing of a pellet as described in the procedure. Do not use microfuge tubes manufactured by Eppendorf Co. for this procedure. Never **dry the DNA pellet** in a vacuum. This is an unnecessary step that can make hydration of the DNA very difficult if not impossible.

Protocol

1. Transfer 400 µl of cells to a 1.5 ml microfuge tube and collect the cells by centrifugation at **room temperature** for 3 minutes. Discard the supernatant. Recentrifuge the cells for a few seconds and remove the remaining medium using a P200 Pipetman outfitted with a capillary micro-tip.

2. Add 400 µl of freshly prepared **solution II** to the cell pellet. Using the flat end of a toothpick break the pellet up into small pieces by swirling the toothpick. Close the tube and mix by inverting four to six times. Place the tube on ice and incubate for 15 minutes.

3. Meanwhile prepare the ammonium acetate–RNase solution. You will need 3 ml of this solution for ten plasmid preparations. Add 3.5 ml of 7.5 M ammonium acetate to a plastic 15 ml conical centrifuge tube and place it on ice. Add 45 µl of RNase A stock solution, close the tube, and mix by inverting two to four times. Place the solution on ice.

4. Add 300 µl (0.75 volume) of ice cold ammonium acetate–RNase solution to each microfuge tube with lysed cells and mix by gently inverting the tube five to eight times. Place the tube back on ice and incubate for 10 minutes.

5. Centrifuge the lysate at **room temperature** for 5 minutes.

6. Pour off the supernatant into a fresh 1.5 ml microfuge tube. Be careful not to transfer the gelatinous pellet.

7. Add 420 µl (0.6 × total volume) of isopropanol to the supernatant. Mix well by inverting the tube three to four times. Incubate at room temperature for exactly 10 minutes.

8. Place the tube in the centrifuge, with the attached side of the lid away from the center of rotation. Centrifuge at maximum speed for 10 minutes at room temperature.

9. Remove the tube from the centrifuge and open the lid. Gently invert and touch the lip of the tube to the rim of an Erlenmeyer flask. Drain the isopropanol. You do not need to remove all of the isopropanol from the tube. Place the tubes back into the centrifuge

in the same orientation as before. **Note:** when pouring off alcohol, do not invert the tube more than once because this may dislodge the pellet.

10. Place the tubes back into the centrifuge in the same orientation as before. **Do not start the centrifuge:** in this step the centrifuge rotor is used as a "tube holder" that keeps the tube at an angle convenient for ethanol washing.

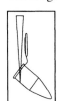

11. Wash the pellet with 700 µl of cold 70 percent ethanol. Holding a P1000 Pipetman vertically, slowly deliver ethanol to the side of the tube opposite the pellet, i.e. the side facing the center of the rotor. Hold the Pipetman as shown in the margin icon. Remove the tube from the centrifuge by holding it by the lid. Gently invert the tube and touch the lip to the rim of an Erlenmeyer flask. Hold the tube to drain the ethanol. You do not need to remove all of the ethanol from the tube. Place the tubes back into the centrifuge in the same orientation as before and wash with 70 percent ethanol one more time. **Note:** this procedure makes it possible to quickly wash a large number of pellets without centrifugation and vortexing. Vortexing and centrifuging the pellet are time-consuming and frequently lead to substantial loss of plasmid DNA.

12. Place the tube back into the centrifuge making sure that the side with the pellet is away from the center of rotation. Without closing the tube lid, start the centrifuge for 2–3 seconds in order to collect the remaining ethanol at the bottom of the tube. The centrifuge should not reach maximum speed. A speed of 500–1,000 r.p.m. is sufficient for collecting all the ethanol remaining on the sides of the tube. Remove all of the ethanol using a P200 Pipetman micropipette outfitted with a capillary tip.

13. Add 15 µl of sterile deionized water to the tube and dissolve the pelleted DNA. Gently pipette the liquid up a down, directing the stream of liquid towards the pellet. **Note:** centrifugation in a microfuge with a fixed angle rotor will deposit most of the DNA pellet on the side of the tube rather than on the bottom. To dissolve all plasmid DNA when these centrifuges are used, it is necessary to direct the stream of the dissolving solution **towards the lower two-thirds of the bottom side of the tube**.

ABI sequencing reactions require at least 1–3 µg of pure plasmid DNA. The concentration of plasmid and its quality can easily be established using mini-gel electrophoresis. Alternatively the concentration of plasmid DNA can be measured by absorbance at 260 nm using a spectrophotometer.

Mini-gel electrophoresis

1. Prepare a mini-gel using a casting tray no larger than 7.5 cm × 7.5 cm and a thin gel (0.2 cm). Seal the ends of the gel-casting tray with tape. Regular labeling tape or electrical insulation tape can be used. Use a mini-gel

well-casting comb with wells 0.2–0.5 cm long and 1 mm (or less) wide. Check that the bottom of the comb is approximately 0.5 mm above the gel bottom. To adjust this height it is most convenient to place a plastic charge card (for example MasterCard) at the bottom of the tray and adjust the comb height to a position where it is easy to remove the card from under the comb.

2. Prepare 500 ml of one times TAE by adding 10 ml of a 50 times TAE stock solution to 490 ml of deionized water.

3. Prepare 1 percent agarose gel. Place 15 ml of the buffer into a 100 ml Erlenmeyer flask and add 150 mg of agarose powder. Melt the agarose by heating the solution in a microwave oven at full power for 1–2 minutes until the agarose is fully dissolved. If evaporation occurs during melting, adjust the volume to 15 ml with deionized water.

4. Cool the agarose solution to approximately 60°C and add 1 µl of ethidium bromide stock solution. Slowly pour the agarose into the casting tray. Remove any air bubbles by trapping them in a 10 ml pipette. Place the comb 1 cm away from one end of the gel. Allow the gel to solidify for 20–30 minutes.

5. Spot 5 µl of TE buffer onto the virgin side of a piece of parafilm. Add 1 µl of stop solution (with dye) to it. Prepare as many "spots" as you have samples. Add 2.0 µl of sample to each aliquot and mix the buffer and plasmid DNA pipetting up and down with a Pipetman. Immediately load the suspension onto the gel. Repeat the above procedure for each sample using a fresh tip for each preparation.

6. Electrophorese the samples at 70–90 V for 25–40 minutes.

7. View the gel with a UV light box and photograph it. Plasmid bands should be visible as strong single or double bands. A typical result of gel electrophoresis of plasmid preparations is shown in Fig. 5.5. If the intensity of the bands of your preparation are similar to those in Fig. 5.5, you will need 1–2 µl of your plasmid preparation for a single sequencing reaction.

Experiment 6: sequencing reactions for an ABI 3700 sequencer

Introduction

In this experiment you will use a modification of dideoxy sequencing called "cycle sequencing." Cycle sequencing uses heat-stable DNA polymerase and a thermal cycler for sequencing. This makes it possible to generate sequence information from a relatively small amount of template. In this modification of PCR amplification, only a single primer (sequencing primer) is used rather than the two primers used in normal PCR reactions. Thus, the amplification is linear rather than exponential.

The newly synthesized DNA will be labeled with dideoxy nucleotides (terminators) that are labeled with fluorescent dye. Each terminator is

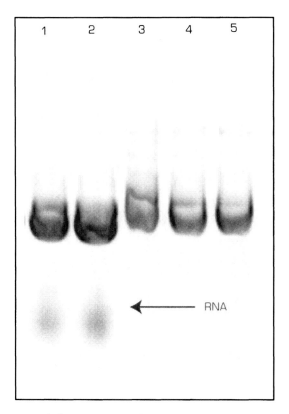

Figure 5.5 Agarose gel electrophoresis of a plasmid preparation. Computer imaging exposure for 1 second. Plasmids were purified from 300 µl of cells grown for 19 hours in TB medium containing 100 µg ml^{-1} ampicillin and 2 µl of plasmid samples were loaded onto a 1 percent agarose gel and run for 25 minutes at 5 V cm^{-1}.

labeled with a different color dye. A color sensor in the automatic sequencer can detect these labels. Sequencing will be carried out using the most modern sequencing machine. It uses a capillary rather than a gel to separate the products of the sequencing reactions. You will sequence three of the best plasmids you prepared in experiment 5.

Technical tips

The thermal cycler for sequencing should be able to accommodate thin-walled 0.2 ml PCR tubes. Since the sequencing reaction volume is very small (10 µl) and use of oil in order to prevent evaporation is not recommended, the thermal cycler used for running sequencing reactions should be equipped with a heated lid. The Idaho Technologies air thermal cycler used in other PCR experiments described in this book is also an excellent instrument for running sequencing reactions. It should be equipped with a thin tube adapter available from Idaho Technologies. The time required for running sequencing reactions using this instrument is shorter than the time necessary for running these reactions in a standard PCR instrument, making it an ideal instrument for class use.

Protocol

1. Label one 1.5 ml microfuge tube RX and place it on ice. Add the reagents shown in Table 5.4 to this tube starting **from the addition of water**.
2. Label three 0.2 ml thin-walled PCR tubes with your group number and tube number (e.g. 1.1, 1.2, and 1.3). Add 8 µl of the reaction mixture prepared in step 1 to each tube.
3. Add 2.0 µl of your first plasmid to the first tube. Mix well by pipetting up and down. Repeat this procedure for tubes 2 and 3, adding the remaining plasmid DNAs. Use a fresh yellow tip for each transfer.
4. Insert the four tubes into the heating block of the thermal cycler. Start the following temperature cycling program: 95°C for 10 seconds, 50°C for 10 seconds, and 60°C for 3.5 minutes. Use 40 cycles.

Table 5.4 Sequencing reaction mix

Ingredient	Single reaction	Four reactions
ABI buffer mix	4.0 µl	16.0 µl
Primer M13 F (3.2 pm µl^{-1})	2.0 µl	8.0 µl
Water	2.0 µl	8.0 µl

Next day

1. Remove the tubes from the thermal cycler and store them in a −20°C freezer.

FOURTH LABORATORY PERIOD

In this laboratory period you will perform the final step in the preparation of the DNA fragment for sequencing. This step is removal of dideoxy terminators from the sequencing reactions prepared in the third laboratory period.

Experiment 7: removing dideoxy terminators

In this experiment you will remove any remaining unincorporated dideoxy nucleotides from your sequencing reactions. Big Dye-labeled terminators (ddNTPs) can contaminate capillaries of the ABI 3700 sequencer and render them unusable. You will use Edge Gel Filtration Cartridges for removing fluorescent dye, which are specially designed for this task. You should carry out this procedure very carefully because even a small amount of fluorescent dye in your preparation will permanently damage the capillary tubes of the DNA sequencer.

Protocol

1. Place the cartridge assembly (cartridge and bottom tube) into a microfuge. Centrifuge for **exactly 2 minutes at 3000 r.p.m. Note:** this is important because a shorter centrifugation time and lower speed will result in an elution volume greater than the input volume.
2. Transfer the cartridge to a clean microfuge tube that will be provided to you. Discard the bottom tube, which should contain some liquid.
3. Holding the cartridge up to the light, transfer the reaction mixture to the top of the cartridge. Carefully dispense the sample directly onto the center of the gel bed at the top of the cartridge without disturbing the gel surface.
4. Close the cap and centrifuge the cartridge for 2 minutes at 3000 r.p.m.
5. Collect the tube containing sample and discard the cartridge. Up to 4 µl of the sample may be lost during sample processing. This is normal. However, if the volume loss is greater than 4 µl this is an indication that you centrifuged your cartridge too fast or for too long.
6. Label your tubes on the lid with numbers that will be given to you by your instructor. Store the tubes at −20°C until samples are ready for sequencing.

References

Birnboim, H.C. (1983) A rapid alkaline extraction method for the isolation of plasmid DNA. *Methods Enzymol.*, **100**, 243–55.

Birnboim, H.C. and Dolly, J. (1979) A rapid alkaline lysis procedure for screening recombinant plasmid DNA. *Nucleic Acids Res.*, **7**, 1513–23.

Holmes, D.S. and Quigley, M. (1981) A rapid boiling method for the preparation of bacterial plasmids. *Anal. Biochem.*, **114**, 193–7.

Hultner, M.I. and Cleaver, J.E. (1994) A bacterial plasmid DNA miniprep using microwave lysis. *BioTechniques*, **16**, 990–4.

Surzycki, S. (2000) *Basic Techniques in Molecular Biology*. Springer-Verlag, Berlin, Heidelberg, and New York.

Vogelstein, B. and Gillespie, D. (1979) Preparative and analytical purification of DNA from agarose. *Proc. Natl Acad. Sci. USA*, **76**, 615–19.

Wang, B., Merva, M., Williams, W.V., and Weiner, D.B. (1995) Large-scale preparation of plasmid DNA by microwave lysis. *BioTechniques*, **18**, 554–5.

CHAPTER 6

Computer Analysis of Sequencing Data

Introduction

The goal of this laboratory is to introduce you to sequence analysis methods. You will analyze DNA sequences obtained by sequencing your genomic DNA fragments. Your first task will be to retrieve the file of raw sequencing data generated by the ABI automated sequencer. These data are stored as sequencing chromatograms that are represented as different color peaks. Each color represents a different base as recorded by a digital camera and read by a base-calling algorithm of the ABI automated sequencer. You will inspect these chromatograms and resolve ambiguities of base call by the sequencer. You also will remove vector sequences and store files in a form suitable for sequence analysis programs.

Next you will carry out sequence analysis using the computer. This analysis will include a search for similar sequences in the GenBank database using the local alignment analysis programs BLAST (Basic Local Alignment Search Tool) and/or FASTA. You will search the human EST (Expressed Sequence Tag) database to determine whether your sequence is expressed. Using BLAST you will search for the presence of short interspersed nuclear elements (SINEs) and long interspersed nuclear elements (LINEs) in your sequence. Single-sequence analysis will consist of a search for direct and inverted repeats using a dot matrix program and a search for GpC islands and restriction enzyme sites.

The exercise will be carried out during one laboratory period.

Background

The base sequence in nucleic acids and amino acid sequences in proteins describe the primary structures of these molecules. Sequence analysis comprises determination of sequence properties and comparison of this sequence with other known sequences. The analysis is carried out using computer programs that implement specific algorithms for describing sequence properties or which carry out comparisons with sequences present in databases.

A sequence analysis is a vast subject that incorporates large numbers of different methods and tools. These include (i) storing sequences and the construction of databases; (ii) a database search for similar sequences; (iii) sequence pair and multiple sequence alignments; (iv) prediction of the secondary structure of RNA; (v) prediction of protein structure and function; (vi) phylogenetic analysis; (vii) gene prediction analysis; and (viii) whole genome analysis and comparison.

The theoretical background of many of the methods and programs is highly complicated and not easily accessible for biologists. Recently, however, some excellent reviews of these subjects have appeared that make it possible for biologists to understand the underlying algorithms and assumptions made and the limitations existing in the application of most of these methods (Brown, 2000; Higgins and Taylor, 2000; Mount, 2001).

In this laboratory we will concentrate on two tasks: the first will be sequence alignment analysis and the second will be analysis of the properties of a single sequence.

Databases and sequence formats

DNA and protein sequences are stored in large databases. Several databases are used for sequence comparison. Table 6.1 presents a list of the main databases. In addition to these databases, there exist a number of specialized databases. These are, for example, a protein structure database, a subset of protein family databases, or databases for each fully sequenced genome. These are very useful when working with specific genomes or protein families. An easy way of accessing sequence databases on the World Wide Web (WWW) is to use ENTREZ, a resource prepared by the National Center for Biotechnology Information at http://www.ncbi.nlm.nih.gov/Entrez/.

The most important databases listed in Table 6.1 are GenBank, EMBL (European Molecular Biology Laboratory), and DDBJ. Each database collects and processes new sequence data and relevant biological information from scientists in their region, e.g. EMBL collects from Europe, GenBank from the USA, and DDBJ from Japan. These databases automatically update each other with the new sequences collected from each region every 24 hours. The result is that they contain exactly the same information, except for any sequences that have been added in the previous 24 hours. This is an important consideration in your choice of database.

The databases listed in Table 6.1 store data in unique formats. These formats are standard ASCII files but, unfortunately, they differ from each other considerably. These differences are important when running sequence analysis software that may or may not recognize a particular file format.

Table 6.1 Sequence databases accessible through the Internet

Database type	Database name	Database address	Description
DNA	GenBank	www.ncbi.nlm.nih.gov/	DNA sequences (USA)
DNA	EMBL	www.ebi.ac.uk/embl/	DNA sequences (Europe)
DNA	DDBJ	www.ddbj.nig.ac.jp	DNA sequences (Japan)
Protein	SwissPort	www.expasy.ch/sprot/sprot-top.html	Highly annotated protein DB
Protein	PIR	www.gergetown.edu	Annotated protein DB
Protein	GenPept	www.ncbi.nlm.nih.gov/Entrez/protein.html	Translation of GenBank
Protein	Genomes	www.ncbi.nlm.nih.gov/Entrez/genome/org.html	Protein sequences by organisms
Protein and DNA	nr	www.ncbi.nlm.nih.gov/BLAST/	Non-redundant database*

*A non-redundant database is a database that has only one copy of a given sequence. A redundant database can have more than one copy of a given sequence. A redundant database is more comprehensive and more likely to contain recently discovered sequences.

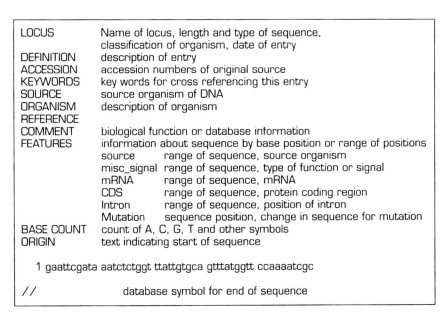

Figure 6.1 GenBank DNA sequence file format.

The most common file formats that are recognized by nearly all sequence analysis programs are the GenBank format, EMBL format, FASTA format, NBRF format and Stanford University Intelligenetic format.

In most of these formats information is organized in fields that are recognized by an identifier word or letter at the beginning of each text line.

The format of the GenBank file is shown in Fig. 6.1. The GenBank format starts from the word LOCUS on the first line of the text. This is followed by a number of fields: DEFINITION, ACCESSION, SOURCE, etc. The word ORIGIN delineates the last line of the text. All lines after this are base or amino acid sequences. The file ends with the sign "//."

The EMBL sequence entry format is similar to the GenBank format and is shown in Fig. 6.2. The identifier words are substituted by two-letter abbreviations of the fields. The first line identifier is ID, which is equivalent to the LOCUS line in the GenBank format. The last line of the text has the identifier SQ and all lines after that are DNA or protein sequences. The symbol for identifying the end of a sequence is "//."

The FASTA sequence format is shown in Fig. 6.3. The first line begins with a ">" character as the identifier. That can be followed by the name or origin of a sequence 60 characters long. No other fields are included in this format. The second line is a line with DNA or amino acid sequences. No spacing or numbering is allowed in the description of the DNA or protein sequences. If two or more sequences are listed in a single file, each sequence is ended by the character "*." The presence of this character may or may not be essential for reading the FASTA format by some sequence analysis

```
ID            identification code for sequence in the database
AC            accession number giving origin of sequence
DT            dates of entry and modification
KW            key cross-reference words for lookup up this entry
OS, OC        source organism
RN, RP, RX, RA, RT, RL literature reference or source
DR            i.d. in other databases
CC            description of biological function
FH, FT        information about sequence by base position or range of positions
              source range of sequence, source organism
              misc-signal range of sequence, type of function or signal
              mRNA range of sequence, mRNA
              CDS range of sequence, protein coding region
              intron range of sequence, position of intron
              mutation sequence position, change in sequence for mutation
SQ            count of A, C, G, T and other symbols
gaattcgata aatetctggt ttattgtgca gtttatggtt ccaaaatcgc cttttgctgt 60

//            symbol to indicate end of sequence
```

Figure 6.2 EMBL sequence file format.

```
>Cp Chloroplast region 2
ACTTGTTGCCATGGTACGTACGTACGGT
TGGCCCATTCGGTACCTGCCATTGCATT*
```

Figure 6.3 FASTA sequence file format.

programs. It is therefore customary to include this character even in files containing a single sequence.

The NBRF sequence entry format is identical to the FASTA format except that the second line contains information about the sequence and the third line contains the sequence.

The Intelligenetic sequence entry format is very similar to the NBRF format except that a semicolon is placed before the first line instead of the ">" character. The second line contains an identifier describing the sequence. The third and subsequent lines contain sequences. A number 1 is placed at the end of the sequence if the sequence is linear or a number 2 if the sequence is circular.

A number of programs exist that will convert one format into another format. The most popular is the READSEQ program developed by Dr Gilbert at Indiana University. The program is available on the Internet at http://dot.imgen.bcm.tmc.edu:9331/seq-util/readseq.html or it can be downloaded from the FTP site ftp.bio.indiana.edu/molbio/readseq.

Sequence alignments

Comparing two or more sequences to each other is called sequence alignment. Sequence alignment is the most important and commonly used method of sequence analysis. The first thing to do with a newly determined sequence is to compare it to all known sequences. The goal is to determine whether this sequence is identical to known sequences or does this sequence have some degree of similarity to known sequences.

Sequence similarity or identity may indicate a similar structure and suggest the function of an unknown sequence. Moreover, finding dissimilar regions in sequences that are otherwise identical is also very important. These dissimilarities can have a different origin, such as population polymorphism, differences in multiple copies of a gene in a single individual, or evolutionary divergence of genes in different organisms. Thus, sequence analysis is a necessary first step in detailed experimental studies of structure, function, evolutionary origin, and relations between biological molecules.

Alternatively, sequence alignment can be used "in reverse," that is one can use a sequence with known function for searching through the sequence database (e.g. whole genome sequence) of a particular organism in order to identify a gene that may have the same function.

The process of sequence alignment involves one-to-one matching of two strings of letters (nucleotides or amino acids) so that each letter in a pair of sequences is associated with a single character of the other sequence or with a null character or gap. As its basis, the process of comparison can be imagined as writing two sequences across a page in two rows in a way that identical characters are placed in the same column and non-identical characters are placed as the gaps or mismatches. An optimal alignment is considered an alignment that places the maximum number of identical characters under each other in both sequences. The alignment of two sequences without gaps requires an algorithm that performs a number of comparisons that are proportional to the square of the sequence length. If an alignment is to include gaps at any position and over any length in each sequence, the number of combinations of gaps and matches, even for two short sequences, becomes very large and it is impossible to find the best alignment by trying all possibilities. It was calculated that the number of comparisons that would be required to compare two sequences with 300 characters would be 10^{88} (Waterman, 1989). In order to realize how big this number is it should be compared to the estimated number of elementary particles in the universe: 10^{80}. The essence of alignment methods (programs) is to solve this problem in a realistic time and to give statistical measurements for the quality of the alignment.

There are several different alignment algorithms used in sequence analysis. Most of these algorithms use a dynamic programming method. The number of combinations in dynamic programming is limited by the following approach. The alignment of one sequence with another is represented as a grid, with each sequence on an axis. Each cell of this matrix ties a pair of units (amino acids or nucleotides) in the two sequences. The best alignment of two sequences is the path from one end of the matrix (upper left

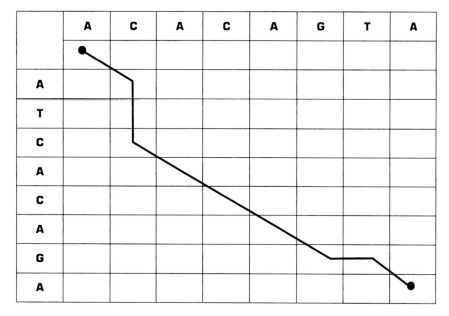

Figure 6.4 Graph of the dynamic programming decision matrix.

corner) to the other (lower right corner) that passes through most matching cells.

Figure 6.4 shows a dynamic programming decision matrix for alignment of two sequences. Wherever sequences are identical, the path moves diagonally. When the sequences differ, the path can move vertically or horizontally, indicating the insertion of gaps in one or another sequence. At each step, the computer chooses the path through the most identical cells of the matrix that has highest score from all previous cells that brought the path to this point. Thus, for any given comparison of two sequences, there can be more than one optimal path, i.e. alignment. It is important to realize that only a few of these optimal alignments may have biological significance and the decision as to which one of them has biological importance cannot be made by the computer.

In order to carry out dynamic programming analysis, an alignment must use a scoring matrix that assigns values for identical scores and mismatched scores and assigns a penalty for gaps. There are several different scoring matrixes in use for nucleic acid and protein alignments. The use of one matrix over another determines the stringency of the alignment, e.g. finding closely related sequences or evolutionary distant sequences, etc. The dynamic programming method was introduced to biology by Needleman and Wunsch (1970) and is usually referred to as the Needleman–Wunsch algorithm. Smith and Waterman (1981a,b) extended and enhanced this method to include an improved scoring system. This algorithm is referred to as the Smith–Waterman algorithm.

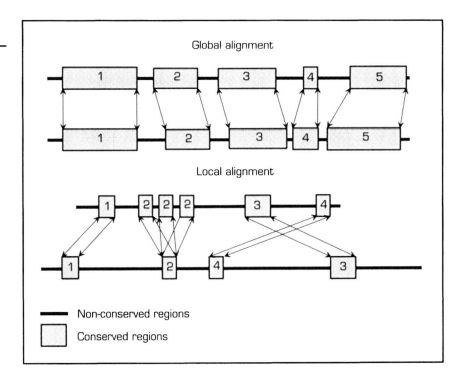

Figure 6.5 Principle of global and local alignments. Two DNA sequences are compared for each alignment. Five regions of similarity are indicated in global alignment. These regions are at approximately the same positions in both sequences. Local alignment of two sequences shows four conserved short regions. They are not at similar positions in both sequences.

There are two types of sequence alignment: global alignment and local alignment. The principle of both alignments is illustrated in Fig. 6.5.

In **global alignment** one attempts to derive an optimal alignment between two sequences over their entire length. Sequences that are similar and approximately the same length are usually compared using this type of alignment. The alignment is implemented by using the Needleman–Wunsch algorithm and finds global similarities between two or more sequences. This type of analysis is not sensitive enough for comparing highly divergent sequences and cannot be used for similarity searches with databases. Its most frequent use is in the construction of evolutionary trees or an analysis of closely related proteins.

Most sequences cannot be compared using global alignment algorithms. This is because, in most cases, the similarities between two sequences are limited to specific short regions or domains. Indeed, most proteins are constructed from a combination of specific "modules" called domains. One can imagine the structure of protein as a building constructed from Lego building blocks (domains). Thus, one can construct an almost infinite number of Lego houses using a small number of building blocks! This modular evolution played a major role in the evolution of most protein and DNA sequences.

In order to analyze these structures the **local alignment** method is used. Optimal alignments are made over short regions of similarity that may exist in two sequences rather than a comparison of their entire length. Thus, conserved regions can be found in two sequences even if most of the sequence is dissimilar. The Smith–Waterman algorithm is used for this type of alignment. Routine database searches are possible using a modification of the Smith–Waterman algorithm, but the searches are approximately 50-fold slower than when a search is carried out with heuristic algorithms (based on a process of successive approximations). Database searches can be carried out on the net using this implementation of the Smith–Waterman algorithm with the program SSEARCH at http://fasta.bioch.virginia.edu/fasta/cgi/searchx.cgi?pgm=fa.

The Smith–Waterman algorithm is not used for everyday database searches because it runs very slowly, particularly when very large databases are searched. Routine searches use heuristic algorithms that are very fast. Heuristic algorithms are not guaranteed to find the optimal alignments and might result in some loss in the rigor of comparison by missing weak similarities or identifying similarities that are biologically irrelevant. It is therefore very important to pay close attention to the statistical significance of the results, understand the options presented for optimizing ones search, and be aware of the limitations of each option.

The most popular heuristic programs for similarity searches are BLAST and FASTA. They differ in sensitivity and speed, BLAST being less sensitive but faster and FASTA being slower but more sensitive. Since BLAST performs a faster search than FASTA, it is usually the first choice for searching large databases. FASTA is used if BLAST searches are not successful or give misleading results.

BLAST

The BLAST algorithm was developed for performing fast similarity searches using very large databases (Altschul et al., 1990, 1994, 1997). Access to the BLAST system is possible through the Internet at http://www.ncbi.nlm.nih.gov/BLAST/. There are also numerous mirror sites that provide a BLAST database search. Since we will be using the BLAST program extensively, it is important to understand how this program works and how its options may affect the results.

In the initial scanning step, BLAST compares a query sequence (your sequence) to each sequence in the database. The algorithm breaks each sequence into short fragments designated as words and then looks for closely matching pairs of words between the two sequences. All matching words with similarity scores exceeding a certain preset threshold are saved. These segment pairs are sequences of the same length, one from each sequence. A score is assigned using a designated scoring matrix (usually BLOSUM62 for proteins and PAM for nucleic acids). The sum score is used for determining the degree of similarity. Sequences with a high score are referred to as **high-scoring segment pairs (HSPs)**. The program extends the best HSPs (those with the highest score, i.e. the best matches) in both directions until the maximum possible score for the extension is reached. Those sequences with higher similarity scores are reported as **MSPs** (maximal-scoring segment pairs). Finally, multiple MSP regions are combined and the statistical significance of the similarity score is calculated using the Poisson or sum statistic (Altschul et al., 1994). The most significant hits and their statistical significance (E value) are reported. The value E describes the number of hits one can "expect" to see just by chance when searching a database of a particular size. It decreases exponentially with the score (S) that is assigned to a match between two sequences. Essentially, the E value describes the random background noise that exists for matches between sequences. Thus, the smaller this number is the more probable that the similarity is not random and has biological significance.

The word size option

One of the most important options in BLAST is the **word size**. The lengths of the word determine a fragment size that must have a perfect match to be extended. The default is 11 for BLASTn (nucleic acid search). The BLAST program will scan the database until it finds words that are 11 letters long exactly matching a word of 11 letters in the query. This match will be extended. The 11 letter word is used as a default because it will exclude even moderately diverged homologs from extension and therefore will exclude almost all chance alignments.

Changing the word size will change the speed of the program execution, as well as its output. A small word size will increase the speed and obtain a large number of short exact matches that might not have biological significance. For example, the BLAST search type "Search for short nearly exact matches" uses a word size value of seven.

The filter option

BLAST version 2.0 enables the application of a filter. The filter masks regions of the query sequence that have low compositional complexity (e.g. *Alu* sequences). Masking is achieved by replacing the sequence with a string of Ns (NNNNNN). N is the IUB (International Union of Biochemistry) code for any DNA base. Only the query sequence is masked. The sequences in the database will not be masked. Filtering is necessary because of the large number of repeated or identical sequences (e.g. poly-A tails, proline-rich sequences, etc.) that are dispersed throughout the genome and, therefore, also throughout the database. They will return artificially high scores and misleading results. By default, filtering is turned on to "low complexity." Filtering eliminates statistically significant but biologically uninteresting reports from the BLAST output. When working with human sequences, one can turn default "human repeats" instead of "low complexity." This option masks human repeats (LINEs and SINEs) and is particularly useful for human sequences that may contain these repeats.

The choose database option

This option will limit the search to a particular database. The available databases are as follows.

nr: all GenBank plus EMBL plus DDBJ plus Protein Data Bank (PDB) sequences (but no expressed sequence tags (ESTs), sequence tagged sites (STSs), genome survey sequences (GSSs), or phase 0, 1, or 2 high throughput genomic sequences (HTGSs). No longer "non-redundant."
month: all new or revised GenBank plus EMBL plus DDBJ plus PDB sequences released in the last 30 days.
Drosophila genome: *Drosophila* genome.
est_others: EST sequences of GenBank plus EMBL plus DDBJ.
est_human: human expressed sequence tags.
est_mouse: mouse expressed sequence tags.
dbsts: STSs from database from Bank plus EMBL plus DDBJ
htgs: unfinished HTGSs.
gss: GSS, includes single-pass genomic data, exon-trapped sequences, and *Alu* polymerase chain reaction sequences.
S. cerevisiae: yeast (*Saccharomyces cerevisiae*) genomic nucleotide sequences.
E. coli: *Escherichia coli* genomic nucleotide sequences.
pdb: sequences derived from the three-dimensional structure from the Brookhaven Protein Data Bank.
vector: vector subset of GenBank.
mito: database of mitochondrial sequences.
alu: select *Alu* repeats from REPBASE, suitable for masking *Alu* repeats from query sequences.
Epd: Eukaryotic Promotor Database.

The expect value option

This value is used as a convenient way of creating a significance threshold for reporting results. When the expect value is increased from the default value of ten, a larger list with more low-scoring hits will be reported. The meaning of ten is that, in a database of the current size, one might expect to see ten matches with a similar score simply by chance.

FASTA

The FASTA algorithm was developed by Lipman and Pearson (1985) and Pearson and Lipman (1988). It uses an algorithm that is similar in concept to a dot plot. Similarly to BLAST, FASTA makes a list of all words in each sequence. The words are called KTUP values and are usually two for amino acids and four to six for nucleotides. Then the program identifies words that are identical between the two sequences and a check is made if these words are located close to other identical words in these pairs of sequences. Only non-overlapping words are counted. The next program tries to join high-scoring words, introducing gaps. Whereas BLAST relies on the sum match probability for each local alignment for the sequence, FASTA scores only exact matches. FASTA allows gapped searches to be made. Like BLAST, FASTA is heuristic, sacrificing some speed for sensitivity.

Sequences as short as ten nucleotides in length can be queried using FASTA. The speed of the alignment is largely determined by the **KTUP** value, which is used to limit the word length. In BLAST, a "word" is a short region of the query sequence that is compared against the database. In FASTA, the word is not scored, but must be an exact match if it is to be processed further.

The FASTA output is essentially very similar to the BLAST output. A list of sequences is presented with the most significant alignments first. The best region of the match is reported as the "initial score" (init1). The optimized score represents the score from joining all scoring regions and applying statistical treatment in order to extend the size of the match. Naturally, if the sequences are related, the optimized score is much higher than the initial score because these sequences will have more than one identity region. Actual base alignments are shown in the context of the database sequence that matches it. The numbers of bases that match exactly are reported as a percentage of identity. In the list of all reported sequences the last value is the score and the last number given on each line is the **expect** value (scoring E value). The maximum (threshold) E value is 2.0 by default. As with BLAST, the smaller the expected value, the lower the probability that the reported alignment is a chance finding. Or expressing it another way, **the lower the reported expected value is for a reported sequence, the more likely it is that it is true.** The expect values should be regarded as guidance tools only for identifying the origin of the query sequence.

BLAST versus FASTA

FASTA and BLAST both perform an identical function – to search databases for similar sequences. The way this is achieved by both programs is quite different. FASTA in general is more sensitive than BLAST but this comes with the penalty of speed because the FASTA search is much slower than the BLAST search. However, the choice of each program does not depend only on its speed. Some other considerations might favor one program over the other. FASTA is more sensitive for DNA–DNA searches, particularly for diverged sequences. Moreover, FASTA is better for finding long regions of similarities. However, BLAST is better for finding short regions of high similarity. In general, any search should always start with the BLAST program, and if the search is negative, FASTA should be used.

Single-sequence analysis

Finding the position of features in a DNA sequence is an important step in establishing its function. This analysis is frequently called single-sequence analysis. Single-sequence analysis of DNA usually involves the following.

1. Analysis of the DNA base composition. Genomes of different organisms vary considerably in their base composition. The base composition of various regions of the same genome can also be very different. Mammalian genomes are organized into large regions of similar base composition called isochores. There are AT-rich isochors, which are usually referred to as paleo-isochores and GC-rich isochors, which are referred to as neo-isochores. The human genome contains five isochores (two paleo-isochores and three neo-isochores) that not only differ in their base composition, but also in their positions on chromosomes and the presence of specific types of genes and repeating elements. Thus, the GC composition of a DNA fragment can point to its position on the chromosome, the presence of SINEs or LINEs, or specific genes (e.g. housekeeping genes).

2. Analysis of the distribution of nucleotide doublets or triplets. Distribution of these features is highly characteristic for particular genomes and is not uniform across a single genome. For example, CpG nucleotide pairs are less common than expected in vertebrates, including humans. The gene-coding sequences usually have a low frequency of CpG. However, first introns and 5′ upstream regions of most human genes have higher than average concentrations of CpG. These regions are called "CpG islands." The presence of a CpG island is considered a good indicator of the presence of coding sequences, i.e. the presence of a gene in close vicinity.

3. Search for sequences coding for proteins. This is the most important task in the analysis of whole genomes, which is frequently referred to as annotation of the genome. Since in higher eucaryotes only 5 percent of the DNA sequence is coding for proteins, identifying these DNA sequences is not a trivial task. The protein-coding regions do have an effect on the composition of the DNA, largely due to three factors: (i) uneven use of an amino acid, since proteins have a restricted range of amino acid composition (e.g. tryptophan is a quite rare amino acid); (ii) uneven numbers of codons for each

amino acid, with this number varying from one to six; and (iii) uneven use of codons. Different organisms and different genes in a single organism have different codon usage. These and other considerations are usually taken in to account when designing genome annotation programs (for a discussion see Mount (2001)).

4. Mapping the positions of various site-specific sequences. These include restriction enzymes recognition sites, promoter sites, ribosome binding sites, regulatory motifs, etc. Specific programs performing these tasks are widely available.

5. Analysis of repetitive sequences. Direct and inverted repeats are common in DNA. In addition, each eucaryotic genome contains a large number of repetitive elements, i.e. tandem repeats and interspersed repeats. These repeats are the subject of many of the experiments described in this book. A genome-wide search for tandemly repeated elements can be carried out using the program Tandem Repeat Finder (Benson, 1999). Searching short DNA sequences for tandem and inverted repeats is usually carried out using dot matrix programs.

Dot matrix analysis

This method is the oldest method of sequence analysis and was first introduced by Gibbs and McIntyre (1970). Later, Maizel and Lenk (1981), Staden (1982), and Pustell and Kafatos (1982) introduced its popular implementations. The method does not align sequences, but it is used for finding similarities. Dot matrix analysis is used for two tasks: (i) detailed pairwise sequence comparison; and (ii) revealing the presence of direct repeats and inverted repeats in DNA.

Pairwise analysis is usually performed using sequences (proteins or nucleic acids) that have been identified by BLAST or FASTA as having some region(s) of similarity to a query sequence. Dot matrix analysis will uncover the exact positions of these regions in the query sequence and find regions of lesser identity that neither BLAST and FASTA can identify.

In a search for repeats the sequence is analyzed against itself and repeats are revealed as diagonal in the plot. RNA folding programs are a specific implementation of a self-analysis dot plot.

What are the basic principles of dot matrix analysis? In order to compare two sequences using this method one sequence is written out vertically with each base (amino acid) representing a row and the second sequence is written out horizontally with each base (amino acid) representing a column. Each letter in a row is compared to each letter in a column and a dot is placed at the corresponding intersection when the letters are identical. The diagonal stretch of dots will indicate regions where analyzed sequences are identical. Direct repeats will show up as diagonal lines of dots and inverted repeats as vertical lines. A break in a diagonal line and its displacement

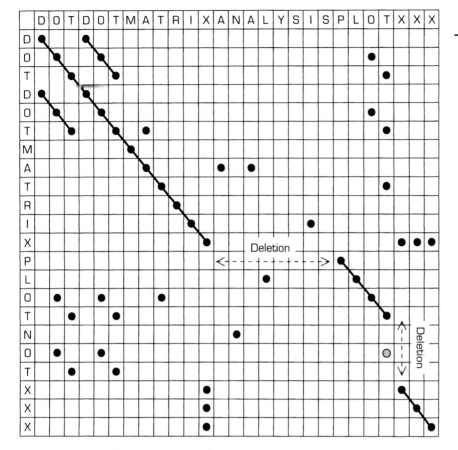

Figure 6.6 Principle of dot matrix analysis. Two "sequences" are compared. The horizontal sequence is DOTDOTMATRIXANALYSISPLOTXXX and the vertical sequence is DOTDOTMATRIXPLOTNOTXXX.

represents a deletion or insertion in one or another sequence. These are common features in pairwise alignment analysis and are important indicators of gap placement. Figure 6.6 illustrates these principles.

Dots that are not on the diagonal will also be present and represent random matches that do not form any significant alignment. This is particularly prevalent in nucleic acid dot plot analysis since these molecules have only four "letters." Various filters are employed for removing this noise. Most popular is the "sliding window" filter. A comparison is made not between individual "letters," but between several of them at the time (e.g. ten). This group is called a window. A dot is only placed if all of the letters or some percentage of them (e.g. 80 percent, i.e. eight bases in a ten-base window) are identical in both sequences. Next the window is moved in both sequences by one or more bases and the comparison is repeated. This process is continued until the entire sequence pair has been analyzed.

This method not only eliminates random dots, but also permits the application of statistics to dot plot analysis as a predetermined percent of the match and the detection of more distant similarities. This can be done by increasing the size of the window and decreasing the percent of matches that will result in the creation of a dot. In more sophisticated dot plot programs, scoring matrix tables are used for determination of an identity between two windows.

Technical tips

The success of sequence analysis depends critically on two steps in the chromatogram-editing task. The first step is to remove poorly sequenced regions at the beginning of the chromatogram (usually ten to 20 bases) and at the end of the chromatogram. Usually sequences after 400 bases are not correct. If both of these sequences are not removed first, it will be very difficult to identify plasmid sequences in the sequence.

Another important step is the removal of plasmid sequences. A plasmid sequence can be present at the beginning of a file (sequences close to the primer) and at the end of an entire sequence. The presence of a plasmid sequence at the end of a file will occur when a fragment inserted into the sequencing plasmid is shorter than 400 bases. Submitting to a BLAST search sequence file with plasmid sequences present will result in a very large output containing all plasmid sequences present in the database. In most chromatogram-editing programs plasmid sequences can be removed automatically. Otherwise they should be removed "by hand."

If a sequence contains tandem repeats or dispersed repeats (LINE or SINE), the output after a BLAST search will also be very large since there are a very large number of these elements in the human genome. In this case, the sequence can be resubmitted for a BLAST search after removing these sequences. LINE and SINE elements are usually located close to protein-coding regions and, thus, their presence can indicate the rest of the sequence codes for proteins.

Removing LINE and SINE sequences from the query file is even more important when searching for the chromosome position of this sequence.

There are a number of non-commercial DNA analysis packages than can be used instead of the Sequencher and DNASIS programs described here. The Staden package can run on Mac and PC computers and can be downloaded from http://www.mrc-lmb.cam.ac.uk/pubseq/staden_home.html. The AnnHyb package for Windows can be downloaded from //annhyb.free.fr/download.php3. DNATools is a shareware package that can be downloaded from www.dnatools.dk. In addition to the usual analysis of a single sequence, this package contains a chromatogram-editing module. Another excellent shareware package for PC computers is "DNA for

Windows" with an excellent module for chromatogram editing. It can be downloaded from http://website.lineone.net/~molbio/.

Most of the packages do not incorporate dot plot analysis. There are several java applets for dot matrix analysis that can be run on any computer. These are as follows.
1. Dottlet at www.isrec.isb-sib.ch/java/dotlet/Dotlet.html.
2. DNA dot at http://arbl.cvmbs.colostate.edu/molkit/dnadot/.
3. DotPlot at http://www.geneart.com/dotplot.php3 (best).

Protocol

Editing chromatograms

The chromatogram editor is used for viewing and editing the raw sequence data produced by automated DNA sequencers. You will do this analysis using the program **Sequencher**. The chromatogram editor displays colored peaks as interpreted by a base-calling algorithm of the ABI sequencer. The chromatogram can be analyzed manually when resolving ambiguities (which may become apparent when assembling sequences) and changes can be made to the derived DNA sequence. In the chromatogram editor (i) sequence data is displayed graphically; (ii) the DNA sequence derived from the chromatogram is freely editable; (iii) each trace can be dragged up or down to help clarify base calls at the beginning and extreme ends of a chromatogram; and (iv) the vertical scale of the chromatogram traces can be adjusted.

1. Download your sequences from the server. The file extension will be abi. Place this file into a folder with your group number. Remove the long header from the file name that was introduced by the ABI sequencer and change it to a file name preferred by you. Do not remove the "abi" extension. Repeat this procedure with each file of your sequences.

2. Open one of the sequencing files by double clicking on it. You will see windows with DNA sequences. Each base is represented by a different color: green = **A**, blue = **C**, black = **G**, and red = **T**.

3. At the beginning and end of your sequence there will be many letter **N**s. This letter is colored light blue. These are the positions in your sequence that the computer could not assign to any specific base (N stands for u**N**known). The beginning and end of the sequence will have the most bases designated N.

4. In the upper right corner of the DNA sequences window, there is a button labeled "**show chromatogram**." Click on it. The window opens showing chromatogram peaks labeled in different colors. On the top of this window, you will see letter designations for each base peak presented in the chromatogram. The base numbers will be indicated also.

5. Inspect the beginning of the chromatogram. Try to correct N to the appropriate base if this is possible. If too many Ns are present in a particular region and they cannot be corrected, you need to remove this entire segment.

6. In order to remove an ambiguous stretch, move the mouse over the **top base letter line**. The cursor will change to a small square. Holding the mouse button down, outline the unreadable segment. The outlined segment should turn light blue. Delete this segment by pushing **delete** on the keyboard. This will delete all the letters (but not the chromatogram) of this region. This letter will also be deleted from the sequence window and the base numbering will be changed. **Be very careful with this deletion, you will not be able to undelete it**.

7. Scan the chromatogram to the end. At approximately position 300 or 400 bases the quality of the chromatogram will start to deteriorate and be nearly unreadable. Base peaks will become wide and flat. The ABI interpreter frequently recognizes this as a long stretch of identical bases (e.g. AAAA). You will need to remove all of these sequences to the end of chromatogram as described in step 6.

8. Next you will inspect other ambiguous positions in your sequence. Move the cursor to each base designated N. Inspect the chromatogram at this position. It is frequently possible to guess the correct base directly from the chromatogram picture. Change N to your guessed base by typing the letter.

9. After all corrections are finished, save the corrected sequence. Click on "Save sequencing project," give it an appropriate name, and click OK.

10. Transfer other sequencing files (files with extension abi) to the window of the sequencing project and correct them as described above. Save the sequencing project.

Removing vector sequences

Sequencing files might contain vector sequences at the beginning or end. These should be removed before the BLAST or FASTA programs can analyze the sequences. In order to screen for vector contamination, you must specify the vector used for amplifying your fragment. You can type this information in yourself or load it from a file. To enter or load vector information perform the following.

1. Choose the **trim vector** command from the **sequence** menu. The vector contamination window will open. Choose the **"choose insertion site now"** button from it. The vector insertion site window will open.

2. This window has two entry fields. The first one allows you to **"load the sites."** The second allows you to **"save your choice."** The program can load vector sequences directly from a file in **VecBase** format.

3. Click the button labeled **"use VecBase file."** The window with vector

names will open. Find the file for the **pUC18** vector and open it. When the polylinker window is displayed, click the site where your fragment was inserted into the vector. Since we cloned our insert into *Sma*I sites, highlight it and click the **OK** button.

4. Save this vector by clicking on the **save sites** button (top of the window). Choose the name for your file in **save as** window (for example, my puc18) and **store it in your folder where the rest of your sequences are located**.

5. Highlight your sequence and choose the "**trim vector**" command from the "**sequence**" file. The computer searches the selected fragments for overlap with the vector bases entered. Any of the searched fragments that are found to contain vector bases will be displayed in the **vector screening** dialog box.

6. The window will show your sequence as a single line in two colors. Blue color indicates your sequence, whereas vector sequences are red. A scissor icon will separate the sequences.

7. Click on the "**show sequence**" button and, instead of a line, you will see base sequences. The *Sma*I site should be displayed on the border between the vector and insert. This site has the sequence CCC|GGG and, thus, you should see three Gs (GGG) in the sequence colored red.

8. Trim unwanted vector sequences by clicking on the "**trim checked items**" button.

9. Perform this analysis with all of your sequences and save the project again.

References

Altschul, S.F., Gish, W., Miller, W., Mayers, E.W., and Lipman, D.J. (1990) Basic local alignment tool. *J. Mol. Biol.*, **215**, 403–10.

Altschul, S.F., Boguski, M.S., Gish, W., and Wooton, J.C. (1994) Issues in searching molecular databases. *Nature Genet.*, **6**, 119–29.

Altschul, S.F., Madden, T.L., Schaffer, A.A., Zhang, J., Zhang, W., Miller, W. et al. (1997) Gapped BLAST and PSI-BLAST: a new generation of protein database search programs. *Nucleic Acids Res.*, **25**, 3389–402.

Benson, G. (1999) Tandem repeats finder: a program to analyze DNA sequences. *Nucleic Acids Res.*, **27**, 573–80.

Brown, M.S. (2000) *Bioinformatics: A Biologist's Guide to Biocomputing and the Internet*. Eaton Publishing, A BioTechniques Book Publication, Natick, MA.

Gibbs, A.J. and McIntyre, G.A. (1970) The diagram, a method for comparing sequences. Its uses with amino acid and nucleotide sequences. *Eur. J. Biochem.*, **16**, 1–11.

Higgins, D. and Taylor, W. (2000) *Bioinformatics: Sequence, Structure and Databanks. A Practical Approach*. Oxford University Press, Oxford.

Lipman, D.J. and Pearson, W.R. (1985) Rapid and sensitive protein similarity search. *Science*, **227**, 1435–41.

Maizel, J.V. and Lenk, R.P. (1981) Enhanced graphic analysis of nucleic acid and protein sequences. *Proc. Natl Acad. Sci. USA*, **78**, 7665–9.

Mount, D.W. (2001) *Bioinformatics. Sequence and Genome Analysis*. Cold Spring Harbor Laboratory Press, Cold Spring Harbor, NY.

Needleman, S.B. and Wunsch, C.D. (1970) A general method applicable to the search for similarities in the amino acid sequence of two proteins. *J. Mol. Biol.*, **48**, 443–53.

Pearson, W.R. and Lipman, D.J. (1988) Improved tools for biological sequence comparison. *Proc. Natl Acad. Sci. USA*, **85**, 2444–8.

Pustell, J. and Kafatos, F.C. (1982) A high speed, high capacity homology matrix: zooming through SV40 and polyoma. *Nucleic Acids Res.*, **10**, 4765–82.

Smith, H.O. and Waterman, M.S. (1981a) Identification of common molecular subsequences. *J. Mol. Biol.*, **147**, 195–7.

Smith, H.O. and Waterman, M.S. (1981b) Comparison of biosequences. *Adv. Appl. Math.*, **2**, 482–9.

Staden, R. (1982) An interactive graphic program for comparing and aligning nucleic acid and amino acid sequences. *Nucleic Acids Res.*, **10**, 2951–61.

Waterman, M.S. (1989) Sequence alignment. In *Mathematical Method for DNA Sequences*, M.S. Waterman (ed.), pp. 53–92. CRC Press, Boca Rotan, FL.

Sequence Alignment with BLAST

BLAST is the alignment program for finding sequences in a database similar to your sequence. Reported alignments (i.e. sequences in the database that show some statistically significant similarity to your sequence) are reported in order of significance. BLAST does not try to match the whole sequence. Look for more details about the BLAST program in the introduction to this chapter.

1. In order to perform a BLAST analysis you first need to copy one of your sequences into computer memory. Open your sequencing project, if it is not yet open. Highlight the file with the corrected DNA sequence and double click on it. The window with your DNA sequence will open. Outline this sequence with the mouse and click on **edit**. In the **edit** window choose **copy selection**. This will copy your sequence (**query sequence**) to computer memory.

2. Open **Internet explorer** and type the address that will open the WWW-based BLAST search www.nci.nlm.nih.gov/BLAST/. Open the BLAST search by clicking on "**standard nucleotide–nucleotide BLAST**."

3. A new page will open with space provided for the query sequence (**search**). Click in this window to move the cursor into it.

4. Copy the sequence in the space provided. Open **edit** and choose **paste**. Your sequence will appear in the search window.

Search of "nr" database

1. Choose the **nr** database in the "**choose database**" window.
2. Start the search with the **BLAST!** button.
3. A window will open that will have **query** = (base number) on the top. The

length of your sequence will be indicated in the brackets under this title, e.g. 246 letters. Change 100 to 50 in the window "**description**." Change 50 to ten in the "**alignment**" window. Click on the "**format**" button to see the results.
4. After a while the "result of BLAST" window will open. It will be updated automatically until the search result is ready.
5. Your sequence will be called **query sequence** by the BLAST program. Scroll down this window until a list of identical sequences appears. The first sequence in the list is the best-matched sequence. Click on this sequence and you will see the exact alignment of the matching region(s).
6. Print the results of the BLAST search. Use the **print** function present in the **file** menu. Choose the page number to print. Print only three to four pages of your results.
7. If one of your sequences has high homology to a sequence in the GenBank, download this sequence to your folder.
8. The name of the file to download is located at the left side of the text and is printed in blue. Click on this name. That will open the window with the GenBank file to which your sequence is homologous.
9. Save this file in your folder. Click on **file** at the top of the window and choose **save as**. **Do not click on the text or save buttons in the window**. The query window will appear. Choose your folder. Next, in the window "**save file as**" give a name to this file. In the **format** window, choose **plain text**. Save the file by clicking on the **save** button.

Search for an *Alu* SINE element

1. Return to the search page and initiate a new search with the same sequence for *Alu* repeats. Change the database from nr to ALU. Proceed with the search as described in the steps for the nr database search. Print the results.

Search for expressed sequences

1. If your sequence is part of an expressed protein sequence it should be present in the human EST database. Return to the search page and initiate a new search using the same sequence. Change the database from the ALU database to the human_EST database. Proceed with the search as described above. Print your results.

Search for the Chromosome Position of the Query Sequence

In order to find the chromosomal position of your sequence we will use a human sequence database located in the Sanger Center. Type the new address in the Internet explorer window: www.ensembl.org/Homo_sapiens/blastview/. Open this site.

1. Click on the "submit a BLAST query" window and pass your sequence to it. Open **edit** in Internet explorer and click **paste**. Your sequence should appear in the window.

2. Initiate a search by clicking on the **search** button. A new window will appear with a "BLAST retrieval ID" box. Read the explanation on how to retrieve BLAST results and follow them.

3. Click on the ID number. A new window will appear. It will have a schematic picture of all human chromosomes. Colored boxes or arrowheads indicate the best scored regions. Red colors indicate the best alignment. Arrowheads (blue or green) indicate other positions of partial homology.

4. Outline the entire picture of chromosomes with the mouse pointer.

5. Print the result. Click on the **file** button of Internet explorer. Choose **print** and the radio button **selected**. Start printing by clicking on the **OK** button.

6. Return to the result window and scroll down to the list of matched sequences. The first number in each row indicates the chromosome number and the next (in red) is the name of the sequence, which is followed by the score value and E value. Click on the name of the sequence. You will see alignment between the query sequence and the sequence in the database. Print this page.

7. Return to the window with the chromosome picture. Move the cursor over the arrowhead of the box with the best alignment (red box). A little window will appear. Click on the **show in Contigview** sign. A new window will appear with a detailed view of the position of your sequence on the chromosome, i.e. its region and band numbers. The positions of nearby known genes and DNA marker sequences (labeled D) are also indicated. Outline the **overview** window with the cursor and print it as described above.

8. Scroll down to the **detailed view** box. It contains a picture representation of details of the protein-coding regions, the positions of exons and introns, etc., located close to your sequence, as well as the positions of known mRNA and a list of homologous proteins from other organisms. This window will also indicate whether your sequence is part of a known or predicted human protein. Move the cursor over any filled rectangle that indicates an exon sequence. First, click on any exon (if it exists) in the line "human proteins." A little window will appear, indicating the name of a protein. Click on the **protein homology** sign. The sequence of the protein will appear. Record this data. Next click on any rectangle in the **protein** row. Click on the **protein homology** sign and the window that lists all homologous proteins from other organisms will appear. Make a note as to what these proteins are and what are their functions.

9. Move to the next step of analysis, single-sequence analysis.

Single-sequence Analysis

You will perform a single-sequence analysis. First you will determine the base frequency of your sequence. Next you will make a restriction enzyme map and perform dot matrix analysis. You will use the DNASIS program for this analysis. This program is available both for Mac and PC platforms. Any other program suites such as Staden or GCG can also be used for this analysis.

Converting file formats

In order to perform single-sequence analysis you need to export your sequences from the **Sequencher** program. We will use the GenBank file format. You will export your sequences to your folder in this format. To do so, follow this procedure.
1. Click on a file to be exported in the Sequencher project window.
2. Open the **file** menu and choose the **import & export** window. Choose **export sequence(s)** from this window. The export window will appear.
3. Choose the folder in this window to which you will export this file. Choose your folder and highlight it. Next, change the extension of your file in the **export as** window from **abi** to **seq**. Do not change the name of your file only its extension (for example, your file name may be myfile.seq).
4. Open the **file format** window and change the file format to the **GenBank** format. Click the select button.
5. Export the file by clicking the **save** button.
6. Print your file. Open your folder with the exported file. Highlight the file that you want to print and drag it to the word-editing program icon (e.g. BBedit for Macintosh computers or Word Edit for PCs). Open the **file** menu and choose **print**.
7. Alternatively, you can convert the file using the READSEQ program. Outline the sequence in the Sequencher window and copy it into computer memory.
8. Open the WWW site html://searchlauncher.bcm.tmc.edu/seq-util/seq-util.html in Internet explorer. Choose the **ReadSeq** button and click on **O** (full option button). In the new window open **edit** and click on **paste**. Your sequence will appear in the window. Choose the format GenBank and click on the **perform conversion** button. Click on **save as**, change the file name, and give it seq extensions (e.g. myfile.seq). Change the file format to text and save it into your folder. You can also print this file using the **print** command.
9. Export all your files following one of the described procedures.

Base content analysis

Open the **DNASIS** program by clicking on its icon. Import your files into this program.
1. Open the **file** menu and click on **open**.
2. Open your folder and click on your file with extension *****.seq**. A note will appear that indicates that the program cannot identify this file type. Click the **open** button anyway. A window will open for file identification. Click on the **DNA** button and click **OK**. A window with your DNA sequence will appear. Make sure that there is no other text incorporated other than the DNA sequences (i.e. only A, C, G, or T) at the beginning of this file.
3. Import all your files into DNASIS following the procedure described.
4. Click on the **function** menu and choose **content** and from the content menu choose **base content**. A window will open to set the parameters for this function. It will contain the name of the file and several parameter settings. Choose **window size** 50 and bases G and C.
5. Click on **go**. A graph of base content will appear. You can increase the size of the window by dragging one of its corners. You can also increase the size of the graph (red) in the window by dragging its corner. **Print this graph**.

Restriction enzyme site analysis

Next you will analyze restriction enzyme sites. Use only one of your files for this analysis. Choose one of the sequences by clicking on it and follow the procedure described below.
1. Open the **function** menu and choose **search**. From the **search** menu, choose **restriction enzymes**. A window will open for setting the parameters of the restriction enzyme search. Do not change any parameters. Click on the **go** button. After a short time, a window will open with results. This window tabulates all of the results. Increase the size of the widow by dragging its corner and inspect the results. You do not need to print them.
2. In the upper left corner of the result window, you can see **several icons**. These icons control how data are presented. The first icon, which depicts **tables**, is marked. Right under it is another icon that depicts **circle/line**. Click on it. You will see a presentation of the data in graphic form. Each enzyme-cutting site is presented as a single line with vertical marks. The number of cuts is also indicated. Print this graph by clicking on **print** in the **file** menu. In the print window that opens select **pages from** 1 to 2. This will print only **two pages**. Close the window.

Dot matrix analysis

In order to perform a dot matrix analysis choose one of your sequences that was imported into the DNASIS program. Choose the sequence by clicking on it and follow the procedure described below.

1. Click on the **function** menu and choose **compare** and click on **homology plot**.
2. A window will appear that controls the parameters of this function. **Vertical sequence** will have the file name that you chose to analyze. **Horizontal file** does not contain any sequence. Click on **file selection** and choose the same file that is listed in **vertical sequence**. Change the **check size** to ten and **matching base** to eight or nine. Press the **go** button.
3. A window will appear displaying the results. You should increase the size of this window by dragging one of it corners. The graph will remain small. You should increase the graph size to the size of the window by dragging the corner of the graph.
4. If present a region of direct repeats will appear as a parallel line to the diagonal lines some distance from it (see Fig. 6.6 for an explanation). The distance of the line from the diagonal indicates the distance between tandem repeats. In order to see more detail of the plot in this region, click on the graph icon located in the upper left corner of the data window. Using the arrow, outline the region of the graph that you want to see in detail (hold the mouse button when you make this outline). Release the button and an enlarged graph in the specified region will appear. If you want to move back to the previous size; double click in any area of the graph.
5. Print the enlarged graph. Click on **file** and choose **print. Choose print page 1 to 1 only**.

CHAPTER 7

Determination of Human Telomere Length

Introduction

The goal of this experiment is to determine the length of your telomeric DNA. Telomeres are specialized terminal elements that are present at the ends of most eucaryotic chromosomes. These structures are composed of tandem repetitive sequences that are associated with specific proteins. The length of a telomeric region is a reflection of the "mitotic clock" of normal somatic cells and is therefore age dependent. The length of a telomeric region is inversely related to age: the younger the person, the longer the telomeric region.

In this experiment you will measure the length of your telomere using Southern blot hybridization. First, you will isolate your genomic DNA and digest it with restriction enzymes that do not recognize the telomeric repeated sequence. The chromosomal DNA will be cut into very small fragments except for the telomeric and subtelomeric regions that together comprise the terminal restriction fragment (TRF). The cleaved DNA will be separated by agarose gel electrophoresis, transferred to membrane, and hybridized to probe containing DNA complementary to the telomeric repeat. The TRF will be detected by exposing the blot to antibody against digoxygenin (DIG) conjugated to alkaline phosphatase and chemiluminescent substrate for alkaline phosphatase. You will calculate the average length of your TRF and compare it to standards of "long" (germ line cells) and "short" (old cells) TRFs.

The analysis of telomere DNA length consists of a number of procedures that will be carried out in six laboratory periods. The essential steps are as follows.

1. In this experiment you will purify DNA from your cheek cells. You will learn a new DNA purification procedure that is very quick and does not require the use of organic solvents. In this procedure nucleic acids are purified from cell lysate by precipitating proteins using a "salting out" procedure. This task is described in experiment 1 and will be performed during the first laboratory period. Next, you will measure the purity and concentration of your DNA. This procedure is described in experiment 2 and will be performed in the second laboratory period.

2. Second, DNA will be digested using two restriction endonuclease enzymes. We will use the *Hinf* I and *Rsa*I enzymes. The *Hinf* I enzyme recognizes GA|NTC sequences and the *Rsa*I enzyme recognizes GT|AC sequences. These sequences do not occur in the TRF region of human telomeres. The DNA fragments resulting from digestion will be separated by size by agarose gel electrophoresis. Experiments 3 and 4 describe these procedures. They will be performed during the second laboratory period.

3. Third, DNA fragments will be transferred to a nylon membrane using the Southern blot technique. The membrane will contain the DNA fragments located in exactly the same positions as they were present in the gel. This procedure will be performed in experiment 5 during the third laboratory period.

4. Fourth, immobilized DNA fragments will be hybridized with labeled DNA probe complementary to telomere repeat sequences and hybridized DNA fragments will be detected by chemiluminescence. The average length of telomeres will be calculated using computer analysis. These procedures will be carried out during the fourth to sixth laboratory periods. Figure 7.1 presents the overall timetable for these experiments.

Background

Most procaryotic chromosomes are circular, thereby permitting their easy replication and DNA damage repair. In contrast, eucaryotic chromosomes, including humans' chromosomes, are linear structures containing a single linear DNA molecule. The main disadvantages of having linear chromosomes are difficulty in replicating the 5′-end of the lagging DNA strand (Levy et al., 1992) and the presence of free DNA ends that are open to degradation by nuclease and/or to end-joining reactions that fuse two free ends (reviewed in McEachern et al. (2000)). In order to obviate these problems the ends of eucaryotic chromosomes have specialized structures called telomeres. Telomeres consist of a tandem GT-rich repeat that, for humans and many other organisms, has the sequence TTAGGG and associated specific proteins (Brown, 1989; Evans, 2002). These structures form a double-stranded DNA loop named a t-loop (Griffith et al., 1999). The very end of telomeric DNA is single stranded and forms a 3′ overhang. This overhang displaces an upstream telomere repeat and is tacked inside the double-stranded DNA at the t-loop junction. Formation of this structure is dependent on the telomere-binding protein TRF2. Thus, linear chromosomes are indeed capped on both ends by a circular DNA structure. Figure 7.2 presents a schematic view of a telomere.

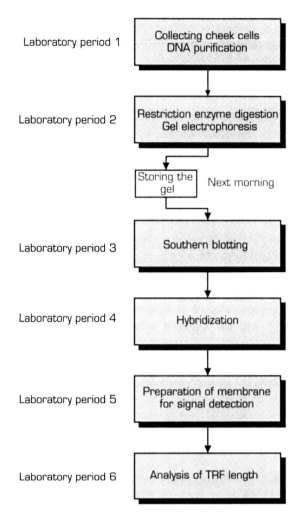

Figure 7.1 Schematic outline of the procedures used in the determination of telomere length.

The length of the telomere region varies considerably among species. The lengths of the simple repeat region can range from 50 bp in *Euplotes* to 300 bp in yeast to over 100 kb in mice. Human telomeres contain 5,000–15,000 nucleotides with a long single-stranded 3′ overhang on at least 80 percent of the telomeres. In addition to the telomere DNA (telomere region), each chromosome contains a subtelomeric region of DNA comprising approximately 4 kb of imperfect tandem-repeated sequences of the original telomeric hexamer.

The length of the telomere region is not stable and depends on cell type and the age of the cell. The telomere region is progressively reduced with each somatic cell division because DNA polymerase α cannot replicate the 5′-end of the lagging strand.

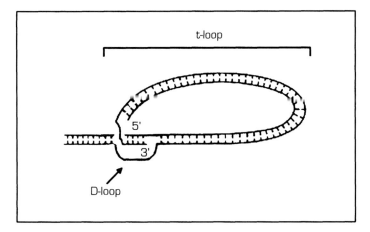

Figure 7.2 Schematic diagram of a telomeric structure on the end of chromosomes.

DNA polymerase α cannot start DNA replication on single-stranded DNA and requires a short double-stranded region with 3'OH in order to start replication (primer). The primer in *in vivo* DNA replication is RNA synthesized by a DNA-dependent RNA polymerase named primase. The RNA primer is subsequently removed, leaving the *de novo*-synthesized strand (and double-stranded DNA) shorter. The next DNA replication will start on the shortened double-stranded DNA and its replication will result in a shorter DNA molecule. The difference in length is the length of the primer. Thus, in every subsequent DNA replication, the chromosome ends are shortened. In humans every DNA replication shortens the chromosome ends by 15–50 bases. Figure 7.3 outlines a schematic of the replication of the ends of linear DNA molecules.

Progressive shortening of the telomere leads to genome instability and aberrant chromosome fusion and rearrangement. Thus, shortening of the telomere region signals the cell to undergo replicative senescence, resulting in cell death. Most normal human cells collected from young individuals can undergo 25 divisions and will enter senescence when their telomeres have been shortened to approximately 5–7 kb. Cells collected from older individuals have much shorter telomeres and will reach a telomere length of 5–7 kb (senescence) after less than 25 cell divisions.

The "end-replication" problem is solved in most eucaryotic cells by activating a specialized reverse transcriptase that is independent of the pre-existing parental DNA template. This enzyme, named telomerase or TERT (**Tel**omere **R**everse **T**ranscriptase), can add a repeat sequence to the 3'-chromosome ends using a short region of its associated RNA moiety as a template (Greider and Blackburn, 1985; Blackburn 1992; Linger and Cech, 1998; Nugent and Lundblad, 1998). Thus, the telomerase rebuilds the ends of chromosomes lost during DNA replication. A schematic diagram of the

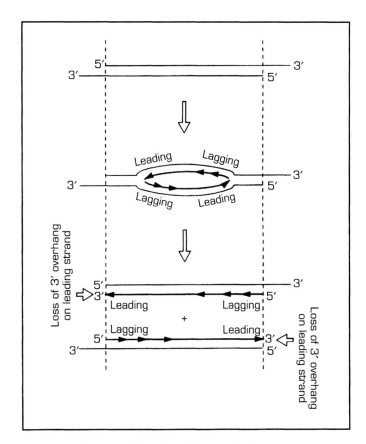

Figure 7.3 DNA replication at the ends of linear molecules.

telomerase reaction that leads to elongation of the single-stranded 3′-end of the telomere is shown in Fig. 7.4.

Telomerase is inactive in most somatic cells and, as a result, the telomere length in these cells is progressively shortened. Thus, the length of the telomere reflects the "replicative history" of the cell, old cells having shorter telomeres than young cells.

Contrary to the situation in somatic cells, in the germ line cell the activity of telomerase is turned on. The restoration of telomere length occurs during the formation of germ line cells. Fusion of gametes creates a zygote with chromosomes containing long telomeres (15–20 kb). Subsequent somatic cell divisions occur without activating telomerase resulting in sequential shortening of telomeres.

Telomerase enzyme is also up-regulated in approximately 85 percent of human tumors leading to continuous division of these cells without senescence (immortalization). Telomerase is activated in most human cancers at the stage when invasive cancer occurs. Thus, during the beginning premalignant stages of cancer development, cells continue to divide in the

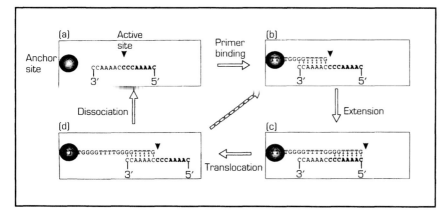

Figure 7.4 Schematic of telomerase activity. Telomerase protein is shown as the rectangle. (a) Active TERT with an RNA template. (b) The 3'-end of DNA binds to an RNA template. (c) Synthesis of 3' DNA by telomerase using RNA as a template. (d) Telomerase translocated to the end of the newly extended 3'-end of the DNA.

absence of telomerase activity. This results in progressive shortening of telomeres, which leads to genome instability and chromosomal rearrangement in these cells. Persistence of this scrambled genome can result in cell death or in the acquisition of mutations that enhance the capability of the tumor. The activation of telomerase at this stage of cancer development reduces further genomic instability and allows these cells to survive and proliferate. As the result of this process, most cancer cells have highly rearranged genomes and very short telomeres.

The significance of telomerase activity to cell aging and cancer has made telomerase a popular subject of research (reviewed in De Lange and DePinho (1999)).

FIRST LABORATORY PERIOD

Experiment 1: isolation of genomic DNA

Introduction

We will be using a genomic DNA purification kit (the Wizard from Promega Co.) that is designed for quick isolation of DNA without the application of organic solvents. DNA purity and yields are not exceptional, but sufficient for the telomere length assay. You will collect your cheek cells and lyse them in a solution containing detergent. Next you will remove the RNA by RNase digestion and precipitate cellular proteins with ammonium acetate (1.8 M). High molecular weight DNA is not precipitated with this salt and is left in the solution. Finally, you will collect and concentrate the genomic DNA by precipitation with isopropyl alcohol. Isopropyl alcohol will be removed by washing pelleted DNA with 70 percent ethanol. You will dissolve your DNA in a small volume of TE buffer and store it for further use.

Background

Removing protein (deproteinization) in the Wizard method is based on the principle of "salting out" of protein, i.e. the phenomenon of decreasing protein solubility at high salt concentrations.

A typical protein in water has hydrophilic and hydrophobic regions (patches) on its surface. The hydrophobic patch forces water molecules to be highly ordered and immobilized when in contact with aqueous solvent. This effectively "freezes" the water around the hydrophobic region. These ordered water molecules are thermodynamically unstable and the water molecules surrounding the hydrophobic patch can be easily removed. When this happens, the hydrophobic regions of proteins interact with each other to form aggregates, causing protein to precipitate out of solution. At high salt concentration, salt ions become solvated, meaning they attract lots of water molecules. When freely available water molecules become scarce the "frozen" water molecules from around hydrophobic patches are pulled off, thereby allowing the hydrophobic regions to interact with other proteins rather than with solvent molecules. Proteins with a large number of or bigger patches will form aggregates sooner and precipitate first, while protein with a few patches will remain in solution even at high salt concentrations. Nucleic acids are fully hydrophilic molecules and cannot be precipitated with salt alone.

In consequence, the deproteinization method used by the Wizard DNA purification kit removes the majority of proteins, in particular those with large numbers of hydrophobic regions but leaves behind highly hydrophilic proteins.

Safety precautions

Each student should work only with his or her own cells. Any student who does not wish to isolate DNA from his or her own cells should be provided with human genomic DNA certified to be free of human immunodeficiency virus. This DNA is commercially available from a number of companies (e.g. Promega Co. and Sigma Co.).

Protocol

Collecting human cheek cells

1. Pour 10 ml of PBS into a 15 ml conical centrifuge tube. Transfer the solution into a paper cup. Pour all the solution into your mouth and swish vigorously for 30–40 seconds. Expel the PBS wash back into the paper cup.
2. Transfer the solution from the paper cup to a 25 ml Corex centrifuge tube and place it on ice.
3. Repeat step 1 one more time with fresh PBS. Expel the mouthwash back into the paper cup and transfer the solution to the same 25 ml Corex tube.
4. Collect the cells by centrifugation at 5,000 r.p.m. for 10 minutes at 4°C.
5. Pour as much supernatant as possible back into the paper cup. Be careful not to disturb the cell pellet. Discard the supernatant from the paper cup into the sink. Invert the Corex centrifuge tube with cells on a paper towel to remove the remaining PBS.

DNA purification

1. Resuspend the cells in 1,000 µl of PBS by gently pipetting the cells up and down. Transfer the cells into a microfuge tube.
2. Centrifuge the cells at maximum speed for 10 seconds in order to pellet them.
3. Remove 700 µl of PBS using a P1000 Pipetman. Resuspend the cells in the remaining 300 µl of PBS by vortexing.
4. Add 600 µl of nucleic lysis solution to the tube. Lyse the cells by pipetting up and down using a blue tip with a cut-off end (P1000). Continue until no visible cell clumps remain. The solution will become viscous indicating that most of the cells have been lysed.
5. Add 3 µl of RNase solution to the lysate and mix the sample by inverting the tube four to five times.
6. Transfer the tube to a 37°C water bath and incubate for 30 minutes.
7. Remove the tube from the water bath and allow the sample to cool to room temperature. This will take approximately 5–10 minutes.
8. Add 200 µl of ammonium acetate solution and mix it with the sample by pipetting up and down using a blue, cut-off tip. The sample should be

uniformly mixed before proceeding to the next step. If the sample cannot be mixed, you can vortex the tube for exactly 20 seconds.

9. Cool the tube on ice for 5 minutes.

10. Centrifuge the tube for 4 minutes at maximum speed. This should precipitate proteins, which will be visible at the bottom of the tube as a tight white pellet.

11. Set a P200 Pipetman to 150 µl and remove the supernatant containing DNA to a fresh microfuge tube. Remember to use a yellow tip with a cut-off end for this procedure. Record the volume of the solution transferred. Be very careful not to disturb the pellet. It is better to leave a small amount of the liquid above the pellet than risk transferring some pellet to the fresh tube.

12. Calculate a 0.6 volume of the solution transferred and add this amount of isopropyl alcohol to the tube. This volume should be approximately 600 µl.

13. Precipitate DNA by inverting the tube several times until the white cotton-like clump of DNA becomes visible.

14. Insert the end of a glass hook into the precipitated DNA and swirl the hook in a circular motion to spool out the DNA. The DNA precipitate will adhere to the hook. **Note:** if at this step DNA does not form a clump and instead it forms several smaller fragments, do not try to collect them on a glass hook as described in step 14. Instead collect precipitated DNA by centrifugation as described in steps 16–20.

15. Transfer the hook with DNA into a microfuge tube filled with 1 ml of 70 percent ethanol. Wash the DNA by gently swirling the glass hook. Pour out the 70 percent ethanol and repeat the wash two more times. Proceed to step 21 to resuspend your DNA.

16. Place the tube in a centrifuge, orienting the attached end of the lid away from the center of rotation (see the icon in the margin). Centrifuge the tube at maximum speed for 5 minutes at room temperature.

17. Remove the tubes from the centrifuge. Pour off the ethanol into an Erlenmeyer flask by holding the tube by the open lid and gently inverting the end. Touch the tube edge to the rim of the flask and drain the ethanol. You do not need to remove all the ethanol from the tube. Return the tubes to the centrifuge in the same orientation as before. **Note:** when pouring off ethanol, do not invert the tube more than once because this could disturb the pellet.

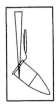

18. Wash the pellet with 700 µl of cold 70 percent ethanol. Holding the P1000 Pipetman vertically (see the icon in the margin) slowly deliver the ethanol to the side of the tube opposite the pellet. Do not start the centrifuge. In this step the centrifuge rotor is used as a "tube holder" that keeps the tube at an angle convenient for ethanol washing. Withdraw the tube from the

centrifuge by holding the tube by the lid. Remove the ethanol as described in step 17. Place the tube back into the centrifuge and wash with 70 percent ethanol one more time.

19. After the last ethanol wash, collect the ethanol remaining on the sides of the tube by centrifugation. Place the tubes back into the centrifuge with the side of the tube containing the pellet facing away from the center of rotation and centrifuge for 2–3 seconds. For this centrifugation you do not need to close the lid of the tube. Remove collected ethanol from the bottom of the tube using a P200 Pipetman equipped with a capillary tip.

20. Add 50 µl of TE buffer to the tube and resuspend the pelleted DNA. Use a yellow tip (P200 Pipetman) with a cut-off end for this procedure. Gently pipette the buffer up and down directing the stream of the buffer towards the pellet. If the pellet does not dissolve in several minutes, place the tube in a 65°C water bath and incubate for 10–20 minutes mixing occasionally. Store the tube in a refrigerator until the next laboratory period.

21. To resuspend the DNA collected on the hook immerse the hook in 50 µl of TE buffer and move the hook back and forth until the DNA is dissolved. Store the tube in a refrigerator until the next laboratory period.

SECOND LABORATORY PERIOD

In this laboratory period you will carry out three experiments. In the first experiment you will determine the concentration and purity of the DNA purified in experiment 1. In the second experiment you will digest this DNA with enzymes that do not digest telomere DNA. We will use a mixture of two restriction enzymes, *Hinf* I and *Rsa*I. We will use a high concentration of the enzyme mixture in order to shorten the digestion time. In the third experiment you will separate TRF DNA fragments from the remaining digested genomic DNA using agarose gel electrophoresis.

Experiment 2: determination of DNA concentration and purity

The theory of measuring DNA concentration and its purity is described in Chapter 1.

Protocol

1. Determine the concentration of DNA by measuring the absorbance at 260 nm. Initially use a 1:20 dilution of the DNA. The absorbance reading should be in the range 0.1–1.5 OD_{260}. Special care must be taken to dilute the viscose solution of DNA when micropipettors are used. Most micropipettes will not measure the volume of a very viscose solution correctly.
2. To prepare a 1:20 dilution of DNA, add 100 µl of PBS to a microfuge tube. Prepare a wide-bore, yellow tip by cutting off 5–6 mm from the end of the tip with a razor blade. Withdraw 5 µl of PBS from the tube and mark the level of the liquid with a marking pen. Using the marked tip, draw DNA solution to the 5 µl mark and transfer it to the tube containing PBS. **Note:** DNA concentration should never be measured in water or TE buffer.
3. Determine the absorbance at 260 nm and calculate the DNA concentration using the formula DNA ($\mu g\,ml^{-1}$) = OD_{260} × 50 × dilution factor.
4. Determine the purity of DNA by measuring the absorbance at 280 nm and 234 nm. Calculate the 260 nm:280 nm and 260 nm:234 nm ratios. Calculate the amount of DNA using equations (1.2) and (1.3) from Chapter 1.
5. Label the tube with your name and group number and indicate the DNA concentration in micrograms per milliliter.

Experiment 3: restriction enzyme digestion

Introduction

In this experiment you will learn how to digest genomic DNA with two restriction enzymes in a single reaction. The enzymes used will be the *Hin*fI and *Rsa*I restriction endonucleases, which do not recognize DNA sequences present in the telomeric and subtelomeric regions of human chromosomes. The simultaneous digestion reaction using these enzymes is possible because both enzymes are active in the same buffer. The buffer that we will use is *Hin*fI enzyme buffer in which the second enzyme retains 100 percent activity.

These enzymes cleave non-telomeric DNA to low molecular weight fragments of less than 500 bp. These DNA fragments will move with the electrophoresis gel front. The digestion will generate telomere DNA fragments of 1–15 kb in size that are well separated from non-telomeric DNA. This makes it easy to obtain good hybridization results with telomeric probes.

You will use DNA isolated from your cells in this experiment. You will also digest two control DNA samples that will be given to you. One of these samples contains DNA with long telomeres and the other DNA with very short telomeres.

Background

The principles and theory of working with restriction enzymes are described in Chapter 2.

Technical tips

The rules for working with restriction enzymes and preparing digestion reactions are as follows.

1. Store restriction endonuclease at −20°C in a freezer that is not frost free at a concentration of $10\,u\,\mu l^{-1}$ or higher.

2. The volume of the digestion reaction should be large enough that the restriction enzyme constitutes no more than 10 percent of the total volume. A 20–30 μl reaction volume should be used.

3. Use a DNA amount no greater than 10 μg added in a volume not to exceed one-third of the reaction volume. Addition of a large volume of DNA dissolved in TE buffer will decrease the Mg^{2+} ion concentration in the reaction, thereby inhibiting restriction enzyme activity. The minimum amount of mammalian genomic DNA that should be used in telomere length analysis is 1.0–1.5 μg per reaction.

4. Use 10–20 units of enzyme per microgram of DNA for telomere analysis. Although this is far more enzyme than is theoretically required, this excess assures complete digestion in the case of impurities in the DNA, decreased enzyme activity from storage, or pipetting errors during enzyme addition. This enzyme concentration will make it possible to complete restriction digestion in 1–2 hours.

Table 7.1 Preparation of restriction reactions

Tube number	Enzyme (10 u µl⁻¹)	Add µl	Ten times buffer	Add µl	DNA type (1 µg)	Add µl	Water (µl)
1H	HinfI/ RsaI	1	Hinf buffer (ten times)	2	Yours		
2H	HinfI/ RsaI	1	Hinf buffer (ten times)	2	Partner's		
3C	HinfI/ RsaI	1	Hinf buffer (ten times)	2	Control (long)		
4C	HinfI/ RsaI	1	Hinf buffer (ten times)	2	Control (short)		

The reactions should be run using a mixture of the *Rsa*I and *Hinf*I enzymes prepared at a ratio of 1 : 1. The buffer used in the reaction should be buffer for the *Hinf* I enzyme (e.g. NEB buffer 2) rather then for the *Rsa*I enzyme (e.g. NEB buffer 1). The *Rsa*I restriction endonuclease retains 100 percent activity in buffer for the *Hinf*I enzyme while the *Hinf*I enzyme has only 70 percent activity in *Rsa*I buffer.

Protocol

1. Label five sterile 1.5 ml microfuge tubes as MRX, 1H, 2H, 3C, and 4C.
2. Calculate the amounts of DNA necessary to add in order to obtain 2–3 µg of DNA per reaction. Use DNA concentrations determined in the previous experiment for your DNA. The concentration of control DNA will be given to you. Record the amounts in Table 7.1.
3. Calculate the appropriate amount of water to add to each tube in order to bring the final volume to 20 µl. Record these values in Table 7.1. Check your calculations again before proceeding further.
4. Start to assemble the reactions by the addition of water. Remember the rule for reaction mixture assembly: the amount of water is always calculated last, but water is always added first.
5. Next add 2 µl of ten times restriction enzyme buffer to each tube.
6. Add the appropriate amount of DNA to each tube. Consult Table 7.1 for the amount and type of DNA to be added. For the addition of each DNA solution you will need to prepare a wide-bore yellow tip. Prepare each tip by cutting off 5–6 mm from the end of the tip with a razor blade. Set a P20 Pipetman to the required volume for each DNA type. Draw that amount of TE buffer into the tip and mark the level of the liquid with a marking pen. Discard the TE buffer. Using the same tip draw the DNA solution to this mark and transfer it to the reaction mixture. Pipette up and down several times to remove the viscose DNA solution from the inside of the pipette tip. **Note:** it is important to follow the procedure described above in order to prevent DNA shearing during pipetting.
7. Start the reactions by the addition of the enzyme mixture. Do not remove the tube containing the enzymes from the ice bucket.

8. Prepare a fresh wide bore yellow tip, as described in step 6 and mix the enzyme with the reaction by pipetting up and down several times followed by a 5–10 second centrifugation.
9. Incubate the reactions for 2 hours in a 37°C water bath.
10. Begin preparing the agarose gel as described in the protocol of experiment 4.
11. After 2 hours of incubation, centrifuge each tube for a few seconds in order to remove condensation from the lid and then stop the reactions by adding 5 μl of stop solution to each digest. Mix well by pipetting up and down with using yellow tip and centrifuge for 5–10 seconds.

Experiment 4: agarose gel electrophoresis

Introduction

In this experiment you will use agarose gel electrophoresis for separating telomeric DNA fragments from fragments generated by digestion of the genomic DNA. Separation will be carried out on a large 0.8 percent agarose gel using TAE (Tris–acetate EDTA) buffer. These gels are well suited for Southern blotting. The genomic DNA fragments are small (approximately 500 bp) and will move ahead of the bromophenol blue dye while telomeric DNA will be located in the middle of the gel. To attain higher resolution, electrophoresis will be run at a low voltage gradient of $1\,V\,cm^{-1}$ overnight.

Background

The theory of agarose gel electrophoresis is described in Chapter 2 of this book.

Safety precautions

Ethidium bromide is a mutagen and suspected carcinogen. Contact with the skin should be avoided. Wear gloves when handling ethidium bromide solution and gels containing ethidium bromide.

For safety purposes, the electrophoresis apparatus should always be placed on the laboratory bench with the positive electrode (red) facing away from the investigator, that is away from the edge of the bench. To avoid electric shock, always disconnect the red (positive) lead first.

Technical tips

The size of the gel for telomeric DNA separation should be approximately 20 cm long, 15 cm wide, and approximately 4 mm thick. To obtain maximum resolution, electrophoresis should be continued until the tracking dye has moved 70–80 percent the length of the gel. The size of the sample well can also affect the resolution of DNA bands. The optimal length of the sample well to use is 1 cm long and 2.0 mm wide. This size of the well will accommodate the whole sample prepared in the digestion experiment. The sample well bottom should be 0.5–1.0 mm above the gel bottom. Most of the commercially available submarine electrophoresis gel boxes fulfill the above requirements.

The molecular weight DNA standard that is suitable for this experiment should have fragments in the range of 1–23 kb. *Hind*III-digested lambda DNA is ideal for this application (e.g. NEB no. 301–2S).

Protocol

1. Seal the opened ends of the gel-casting tray with tape. Regular labeling tape or electrical insulation tape can be used. Check that the teeth of the comb are approximately 0.5 mm above the gel bottom. To adjust this height, it is most convenient to place a plastic charge card (e.g. MasterCard) under the comb and adjust the comb height to a position where the card is easily removed from under the comb.

2. Prepare 1,500 ml of one times TAE by adding 30 ml of a 50 times TAE stock solution to a final volume of 1,500 ml of deionized water.

3. Place 150 ml of the buffer into a 500 ml flask and add the appropriate amount of agarose Weigh 1.2 g of agarose for a 0.8 percent agarose gel. Melt the agarose by heating the solution in a microwave oven at full power for approximately 3 minutes. Carefully swirl the agarose solution to ensure that the agarose is dissolved, that is no agarose particles are visible. If evaporation occurs during melting, adjust the volume to 150 ml with deionized water.

4. Cool the agarose solution to approximately 60°C and add 5 µl of ethidium bromide stock solution. Slowly pour the agarose into the gel-casting tray. Remove any air bubbles by trapping them in a 10 ml pipette.

5. Position the comb approximately 1.5 cm from the edge of the gel. Let the agarose solidify for approximately 20–30 minutes. After the agarose has solidified remove the comb with a gentle back and forth motion, taking care not to tear the gel.

6. Remove the tape from the ends of the gel-casting tray and place the tray on the central supporting platform of the gel box. For safety purposes, the electrophoresis apparatus should always be placed on the laboratory bench with the positive electrode (red) facing away from the investigator, that is away from the edge of the bench.

7. Add electrophoresis buffer to the buffer chamber until it reaches a level of 0.5–1 cm above the surface of the gel.

8. Load the samples into the wells using a yellow tip. Place the tip under the surface of the electrophoresis buffer just above the well. Expel the

sample slowly, allowing it to sink to the bottom of the well. Take care not to spill the sample into a neighboring well. During sample loading, it is very important to avoid placing the end of the tip into the sample well or touching the edge of the well with the tip. This can damage the well resulting in uneven or smeared bands. **Note:** samples must be loaded in sequential sample wells. When loading fewer samples than the number of wells it is preferable to leave the wells nearest the edge of the gel empty.

9. First load 8 μl of the lambda DNA standard into the first well. This standard will be given to you ready to be loaded onto the gel. Next load the entire sample (35 μl) using a P200 Pipetman. Load the samples in the following order: lambda DNA standard, 1H, 2H, 3C, 4C, and lambda DNA standard. **Attention:** standard DNA is loaded on both sites of the samples set.

10. Place the lid on the gel box and connect the electrodes. DNA will travel towards the positive (red) electrode positioned away from the edge of the laboratory bench. Turn on the power supply. Adjust the voltage to approximately $1\,V\,cm^{-1}$. For example, if the distance between electrodes (not the gel length) is 40 cm, in order to obtain a field strength of $1\,V\,cm^{-1}$, the voltage should be set to 40 V.

11. Continue electrophoresis until the tracking dye moves at least half of the gel length. It will take the tracking dye approximately 17 hours to reach this position on a gel 20 cm long.

Next day

1. Turn the power supply off and disconnect the positive (red) lead from the power supply. Remove the gel from the electrophoresis chamber. To avoid electric shock always disconnect the red (positive) lead first.

2. Wrap the gel-casting tray with saran wrap and store in a 4°C refrigerator. Gels can be stored this way for 2–4 days.

THIRD LABORATORY PERIOD

Experiment 5: Southern transfer

Introduction

In Southern transfer, DNA fragments are transferred from the gel to the membrane used for hybridization with a probe. After electrophoresis, the DNA is denatured, transferred to a nylon membrane, and immobilized. The transfer method preserves the separation pattern of DNA on the membrane. Before hybridization DNA will be fixed to the membrane by UV irradiation. The membrane is subsequently used for hybridization with a DIG-labeled telomere probe. DNA of small sizes can easily be transferred directly to the membrane after denaturation. However, large DNA fragments, those bigger than several kilobases, will not transfer efficiently. For this reason, the procedure involves the breakage of large DNA fragments *in situ* before the transfer.

In this experiment you will be using a capillary downward transfer system, which is sometimes called TurboBlotting. This transfer system offers greater speed, target resolution, and convenience than the traditional upward capillary blotting procedure that you learned in the DNA fingerprinting experiment.

The method takes advantage of gravity for accelerating the flow of the transfer solution. Moreover, the downward transfer technique eliminates the need for heavy weights on top of the capillary stack, thus eliminating gel compression that can prevent efficient transfer. The time of transfer can be shortened from 17 hours to approximately 2 hours without lost of transfer efficiency.

Background

The principles and methods of the Southern blotting technique are described in Chapter 2.

Safety precautions

The agarose gel contains ethidium bromide, which is a mutagen and suspected carcinogen. Students should wear gloves when handling these gels. Only powder-free gloves should be used because the procedure uses chemiluminescence for the detection of hybridization. The presence of talcum powder will result in the formation of a "spotted" background. Discard the used gel into the designated container.

When viewing and photographing the gel with a UV transilluminator, gloves, UV-protective glasses, and a face mask should be used all times.

Technical tips

The gel can be photographed using Polaroid film type 667 (ASA 3000) or using a computer imaging system. However, measurement of the telomere length can be made without a photographic image of the gel.

The use of chemiluminescent detection requires a positively charged nylon membrane. These membranes have very low background with chemiluminescent detection. Some producers have developed a chemically optimized, positively charged nylon membrane for chemiluminescent detection. Evaluation of all of the types of commercially available, positively charged nylon membranes showed that the best membrane for chemiluminescent detection is the Magna Graph membrane manufactured by Osmonics Co (Surzycki, 2000).

Instead of the capillary downward transfer system (TurboBlott), the standard upward capillary transfer can be used without seriously affecting the quality of the transfer. The time required for this transfer is 17 hours (overnight) and disassembling of the blot and cross-linking of DNA to the membrane should be done the next day. For overnight blotting, follow the technique described in Chapter 2.

The best signal to noise ratio for the Magna Graph membrane is achieved when DNA is cross-linked by UV light irradiation on both sides of the membrane. Alternatively, the membrane can be wrapped in aluminum foil and baked in an oven at 80°C for 1 hour. The baking step immobilizes DNA on the membrane. Membranes can be stored at room temperature practically indefinitely.

Protocol

1. Transfer the gel to a glass Pyrex dish and trim away any unused areas of the gel with a scalpel. Cut off the gel below the bromophenol blue dye. This part of the gel contains small fragments of genomic DNA. Cut the lower corner of the gel at the bottom of the lane with the first size standards. This will provide a mark for orienting the hybridized bands on the membrane with the bands in the gel.

2. Transfer the gel to a UV transilluminator. Place an acetate sheet on the top of the gel and draw an outline of the gel with a felt-tip pen. Mark the positions of the wells and the position of the cut corner. It is very important that your drawing be as precise as possible. Label the contents of each well on the acetate sheet and mark the bottom left corner of the gel (under well number 1). This will help you locate the positions of the hybridization signals in your Southern blots. Turn on the transilluminator and mark the positions of standard DNA bands on the acetate sheet.

3. Photograph the gel. Use a setting of 1 second at F8 of Polaroid 667 film. One can also use a computer-imaging system to record the results.

Note: this step can be omitted since determination of telomere length does not require photographic imaging of the gel.

4. Transfer the gel back to the Pyrex dish and add enough 0.25 N HCl to allow the gel to move freely in the solution. This will take approximately 150–200 ml of solution for a standard gel size.
5. Place the dish on an orbital shaker and incubate for exactly 10 minutes rotating at 10–20 r.p.m. Decant the acid carefully.
6. Briefly rinse the gel in water (10–20 seconds) and proceed immediately to the next step.
7. Add 200 ml of denaturing solution to the dish and incubate for 20 minutes with gentle agitation.
8. Decant the denaturing solution, holding the gel with the palm of your hand and repeat steps 7 and 8.
9. Rinse the gel once in water to remove most of the denaturing solution trapped on the surface of the gel.
10. Add 100–200 ml of neutralization solution to the dish and treat the gel for 20 minutes with gentle agitation.
11. Discard neutralization solution and repeat step 10.
12. While the gel is being treated, prepare the nylon membrane for transfer. Cut three sheets of Whatman 3MM paper to the size of the gel. Use the drawing of the gel prepared at step 2 as a guide. Then cut the nylon membrane using the 3MM paper cut out as the template. Use gloves and only touch the edges of the membrane. Immerse the membrane in 20 times SSC for 2–3 minutes.
13. Set up the blot. Refer to Fig. 7.5 when setting up the TurboBlotter System.
14. Place the transfer device stack tray on the bench making sure it is leveled.
15. Cut five pieces of Whatman 3MM paper to the size of the GB004 blotting paper provided to you. Place 20 sheets of dry GB004 blotting paper in the stack tray.
16. Dip one sheet of Whatman 3MM paper (prepared in step 15) in 20 times SSC solution and place on the stack of dry GB004 papers.
17. Place the wet transfer membrane on the stack. Remove any air bubbles trapped between the membrane and wet Whatman 3MM paper.
18. Cut away and remove the gel above the wells. Place the agarose gel with **sample wells up** on top of the membrane. Make sure there are no air bubbles between the gel and the membrane. **Note:** because the gel is thinner in the well area, this part of the gel must be removed because the transfer solution may pass preferentially through this part of the gel causing uneven DNA transfer.
19. Wet three sheets of the Whatman 3MM paper prepared in step 15 in 20 times SSC. Wet the top surface of the gel with 2–5 ml of 20 times SSC and place wet Whatman 3MM paper on top of the gel.

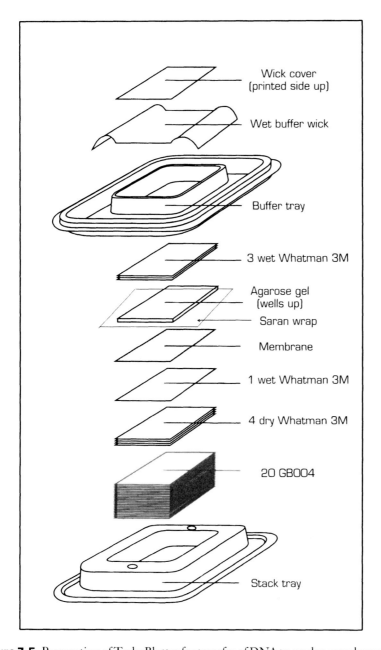

Figure 7.5 Preparation of TurboBlotter for transfer of DNA to a nylon membrane.

20. Cover the entire surface of the gel and surrounding area with saran wrap. With a razor blade "cut away" the saran wrap covering the gel itself. This will leave an opening over the gel.

21. Attach the "buffer tray" of the transfer device to the bottom tray using the circular alignment buttons to align both trays. Fill the buffer tray with 125 ml of 20 times SSC.

22. Start the transfer by connecting the gel stack with the buffer tray using the pre-cut "buffer wick" (included in each blotter stack) pre-soaked in 20 times SSC. Place the wick across the stack so that the short dimension of the wick completely covers the blotting stack and both ends extend into the buffer tray. Place the "wick cover" on top of the stack to prevent evaporation. Make sure the edges of the wick are immersed in the transfer buffer.

23. Continue the transfer for 3 hours. Additional transfer time may be required for gels thicker than 4mm or larger sized nucleic acids. **Note:** if necessary or for convenience this transfer can be carried out overnight. In this case steps 24–26 should be performed the next morning.

24. Disassemble the blot. Using forceps remove the membrane and place it "DNA side" up (the side that contacted the gel) on a clean sheet of Whatman paper. Write your group number with a pencil on the corner of the membrane.

25. Place the membrane on a sheet of dry Whatman 3MM paper. Do not allow the membrane to dry at any time. Place the membrane into a UV oven. Irradiate the damp membrane to immobilize DNA using the automatic setting of the UV oven. Irradiate both sides of the membrane.

26. Store the membrane at room temperature in a plastic bag.

FOURTH LABORATORY PERIOD

Experiment 6: DNA hybridization

Introduction

In this experiment you will hybridize DIG-labeled probe that recognizes telomeric and subtelomeric DNA to the membranes prepared in experiment 5. Probe will be given to you. Hybridized probe will be detected with DIG-specific antibody covalently coupled to alkaline phosphatase. The immobilized probe–antibody complex on the membrane will be visualized using chemiluminescent substrate (CDP-Star). Dephosphorylation of the substrate by the enzyme results in the emission of light, which will be captured on photographic film.

Hybridization will be carried out using a hybridization oven in specially designed hybridization roller bottles. The hybridization procedure uses lower stringency conditions that are necessary for hybridizing short tandem repeats of the telomeric probe. The hybridization time is also substantially shorter than in the standard hybridization procedure. This is because telomeric repeat sequences are present at high concentrations and the complexity of the probe is low.

Background

The theory of hybridization is described in Chapter 2.

Technical tips

Hybridization can be performed in glass or plastic dishes or in sealed plastic bags rather than in a hybridization oven and large roller bottles. Using a hybridization oven is the most convenient method for a large class. No more than one membrane should be placed into the roller bottle. Some membrane overlapping will not affect hybridization results.

Using dishes rather then hybridization bottles will substantially increase the volume of reagents used in each step. When using dishes, the minimum volume of the hybridization solution should be 0.04 ml solution per 1 cm^2 of membrane surface area. The volume of all other solutions should be approximately four to five times larger than for hybridization. Care should be taken that the membrane is sufficiently covered with solutions at all the times and that it can float freely in the container.

One of the great advantages of the DIG system is the stability of the labeled probe. The probe can be stored for an indefinite time at –20°C. It is important to collect hybridization solution with probe after hybridization and store it at −20°C. This probe solution can be reused at least three to four times.

Protocol

Pre-hybridization and hybridization

1. Place the dry membrane into a roller bottle. Make sure that the side of the membrane with DNA is facing away from the glass.
2. Pour 10 ml of Dig Easy solution into a 15 ml plastic centrifuge tube and prewarm it for 10 minutes in a 42°C water bath. Pour it as fast as possible into the roller bottle with a membrane. Close the bottle tightly and label it with your group number.
3. Place the roller bottle into the hybridization oven and allow it to rotate at a slow speed (2–4 r.p.m.) for 30 minutes at 42°C.
4. Ten minutes before the end of pre-hybridization, begin to prepare the probe for hybridization. Add 10 ml of Dig Easy solution into a 15 ml conical centrifuge tube. Add 2 µl of DIG-labeled telomere probe, close the tube tightly, and invert it three to four times to mix the probe with Dig Easy. Place the tube into a 42°C water bath and incubate for 10 minutes.
5. Retrieve the roller bottle from the hybridization oven. Open the roller bottle and pour the pre-hybridization solution into a storage bottle. Pre-hybridization solution can be stored and used again.
6. Remove the tube with the probe from the 42°C water bath and add it to the roller bottle. Return the roller bottle to the oven and allow it to rotate slowly at 42°C for 2 hours.
7. Retrieve your roller bottle from the hybridization oven and pour off the hybridization solution into a 15 ml centrifuge tube. The probe can be stored in a −20°C freezer and reused three to four times.
8. Add 20 ml of washing solution II (two times SSC and 0.1 percent sodium deodecyl sulfate (SDS)) to the roller bottle. Place it into the hybridization oven and allow it to rotate at slow speed at room temperature until the next laboratory period.

FIFTH LABORATORY PERIOD

In this laboratory period, you will continue the hybridization experiment. First, you will remove mismatched hybrids from the membrane using a washing procedure. Second you will prepare your membrane for signal detection by chemiluminescence.

Protocol

Washing reaction

1. Retrieve your roller bottle from the hybridization oven and discard solution II.
2. Add 20 ml of fresh washing solution II to the bottle and place it back into the hybridization oven. Rotate it at maximum speed for 5 minutes.
3. Remove the roller bottle from the hybridization oven and pour off and discard solution II. Drain the liquid well by placing the bottle on end on a paper towel for 1 minute.
4. Add 20 ml of washing solution III prewarmed to 50°C. Place the roller bottle into the hybridization oven preheated to 50°C. Rotate it at a slow speed for 20 minutes.
5. Pour off solution III and discard it. Drain the solution well by placing the roller bottle on end on a paper towel for 1 minute. Repeat the washing with solution III one more time.
6. Remove solution III and drain the roller bottle well as described above.

Preparation of membrane for detection

1. Add 20 ml of buffer A (washing buffer) to the tube. Cool the oven to room temperature and rotate the roller bottle at maximum speed for 2–5 minutes.
2. Retrieve the roller bottle from the oven and discard buffer A. Add 10 ml of blocking solution (buffer B) to the tube. Incubate for 30 minutes, rotating slowly at room temperature.
3. Pour off and discard buffer B. Invert the roller bottle over a paper towel and let it drain well for 1 minute.
4. Add 10 ml of blocking buffer (buffer B) to a plastic conical centrifuge tube and add 1 µl of anti-DIG–alkaline phosphatase solution (antibody solution). Mix well and add the solution to the roller bottle with the membrane.
5. Incubate for 30 minutes, rotating slowly at room temperature. **Note:** the working antibody solution is stable for approximately 12 hours at 4°C.

Do not prolong the incubation with antibody over 30 minutes. This will result in high background.

6. Pour off and discard the antibody solution. Drain the liquid well by placing the roller bottle on end on a paper towel for 2 minutes.

7. Add 30 ml of buffer A to the roller bottle and wash the membrane at room temperature, rotating at slow speed for 15 minutes. Repeat this wash one more time.

8. Retrieve the roller bottle from the oven and move the membrane towards the tube opening by gently shaking the bottle. Do not open the tube for this operation. Open the tube and discard buffer A. Remove the membrane to a Pyrex dish placed on a rotary shaker and add 200 ml of buffer C. Wash the membrane, rotating it slowly for 25 minutes. Discard buffer C and repeat the wash one more time.

9. Place a plastic bag on a sheet of Whatman 3MM paper and add 10 ml of buffer C. Wearing gloves transfer the membrane from the Pyrex dish into the plastic bag. Open the bag and insert the membrane into it. Place the membrane into the pool of buffer and move it with your fingers to the end of the bag. Leave as little space as possible between the membrane and the end of the bag. This will limit the amount of expensive chemiluminescent substrate necessary for filling the bag. Pour off buffer C from the bag. Remove the remaining liquid from the bag by gently pressing it out with a Kimwipe tissue. **Note:** do not press strongly on the membrane because this will increase the background. Most of the liquid should be removed from the bag, leaving the membrane slightly wet. At this time, a very small amount of liquid will be visible at the edge of the membrane.

10. Open the end of the bag slightly, leaving the membrane side that does not contain DNA attached to the side of the bag. Add 0.9–1 ml of CDP-Star solution directing the stream towards the side of the bag. Do not add solution directly onto the membrane.

11. Place the bag on a sheet of Whatman 3MM paper with the DNA side up and distribute the liquid over the surface of the membrane by gently moving the liquid around with a Kimwipe tissue. Make sure that the entire membrane is evenly covered. Do not press on the membrane because this will cause "press marks" on the film. Gently remove excess CDP-Star from the bag by guiding excess solution towards the open end of the bag and onto the Whatman paper with a Kimwipe. Make sure that the membrane remains damp. **Note:** at this point small liquid droplets will be visible on the edge of the membrane, but liquid should not be present on the membrane surface. Seal the bag with a heat sealer.

12. Place the bag in an X-ray film cassette, DNA side of the membrane up. In a darkroom, place X-ray film over the membrane. Expose the film for 5–10 minutes at room temperature. Open the cassette and develop the film using standard procedures for film development. **Note:** maximum light emission for CDP-Star is reached in 20–30 minutes, the light emission

remains constant for approximately 24 hours, and the blot can be exposed to film a number of times during this period. The best results are usually obtained when the membrane is exposed the next day.

13. After exposure, store the bag with membrane at 4°C. The membrane can be stored this way for several weeks.

SIXTH LABORATORY PERIOD

Experiment 7: analysis of TRF length

Introduction

In this laboratory period you will analyze your hybridization autogram. You will determine the mean length of your telomeres, the most frequently occurring telomere length, and the variability in length of your telomeres set. The film will be scanned and image transferred into a computer. The image will be analyzed manually or using the program Telometric.

Background

The telomere length in human cells can vary over one order of magnitude. This is because not all of the cells that you collected are at the same age and because there is considerable heterogeneity of the telomere length even for cells of the same age. The autogram of the telomere sequences or more precisely of the TRF sequences will not show a single band but a number of overlapping length bands forming a smear. Thus, analyzing the TRF length of a population of cells will provide the average TRF length of this sample but not the length of individual TRFs.

The average TRF length can be estimated using an autogram by visually comparing the size of the signal smear to molecular weight markers. However, this is a very imprecise method, mostly because we assume that the most intense part of the smear represents the most frequent band sizes of the TRFs. This is not always true because even a few very long TRF repeats will generate a strong hybridization signal since they have many copies of the same repeat. The precise determination of the size and distribution of the telomere cannot be carried out using the distribution of the intensities of hybridization of the probe having a single repeat element. The distribution of densities should be converted to the distribution of copy numbers, which is not necessarily proportional to film densities.

There are two ways to correct for this. The first method is to integrate the signal intensity over the entire TRF distribution as a function of TRF length. This can be done using the equation

$$L_{TRF} = \frac{\sum (OD_i \times L_i)}{\sum OD_i} \tag{7.1}$$

where OD_i is the chemiluminescent signal intensity and L_i is the length of the TRF fragment (in base pairs) at position i on the gel image, respectively.

This calculation takes into account the higher signal intensity from larger TRF fragments resulting from multiple hybridization of the telomeric-specific probe.

The second method is more sophisticated since it converts the signal intensity at a given point to the relative copy number by adjusting for the number of probes bound to the DNA. This method was recently developed and is performed using the computer program Telometric (Grant et al., 2001). The relative copy number for each DNA is calculated by the equation

$$C_i = \frac{I_i - B_i}{L_i} \qquad (7.2)$$

where C_i is the relative copy number across single autographic line i, I_i is the integrated signal intensity in line i, L_i is the DNA molecular weight (kilobases) in line i, and B_i is the background of the autogram, i.e. the intensity at the position on autogram with no DNA present.

The program generates a plot of the relative copy number of repeats versus molecular weight. This gives a transparent presentation of the actual distribution of the telomere lengths. From this graph a statistic is generated giving the mean, median, and mode of the molecular weight for the telomeric regions. In addition, the variance and semi-interquartile range (heterogeneity for asymmetric distribution) is also calculated.

Technical tips

In order to calculate telomere length successfully, the exposure time of the film should be as short as possible in order to create an autogram having densities of the exposure in the linear range of the film.

Manual determination of telomere length follows the recommendation given in the Roche instruction manual for the Telo TAGGG kit (Roche Molecular Biochemicals Co., 1999). This technique can be carried out on any Wintel or Mac platform. The image should be scanned and transferred to a graphic program capable of measuring the density of the image. The best free program to use is NIH Image, which can run on PC and Macintosh computers. The Macintosh version can be downloaded from //rsb.info.nih.gov/nih-image/ and the PC version (Windows 98 and up) from www.scioncorp.com.

The Telometric program can only run on Macintosh computers and is written as a macro for the NIH Image program. The macro can be downloaded from www.biotechniques.com. It should be installed into the NIH program before use.

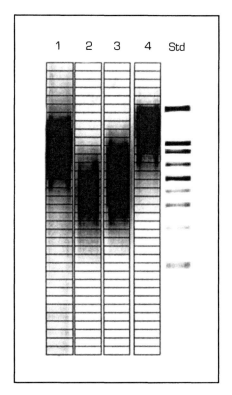

Figure 7.6 Determination of the average length of the TRF region using the manual procedure.

Protocol

Manual method

1. Align the X-ray film with the gel drawing made previously. The outline of the filter should be visible on the film. Transfer the positions of DNA standard bands onto the film. Measure the electrophoretic mobility of each standard band and plot it as a function of the log of fragment size (in kilobases) versus distance traveled. This standard plot will be used to determine the molecular weight of an unknown sample.
2. Divide the scanned image into a grid of columns consisting of each gel lane and rows that cover the entire vertical length of the image. The rows should be 0.5 cm long. See Fig. 7.6 for clarification.
3. Scan the autoradiogram with a computer scanner and save the image into your folder on the hard disk.
4. Determine the size of the TRF present in each square using the standard curve prepared in step 1. The TRF size is taken as the molecular weight (in kilobases) at the middle of each square. Record the size for each square in a table. An example of such a table is shown in Table 7.2.

Table 7.2 Calculation of the TRF average length

Square number	Sample number (line)	OD (mean)	L (size in kb)	OD × L
1	1	185	12.0	2,220
2	1	200	9.0	1,800
3	1	210	6.0	1,260

5. Open the program NIH Image. Import your file into this program. Click on the **file** and **import** buttons. Select your file name and click on it. The image of your autograph with labeled squares should appear on the desktop.
6. Determine the density (OD) for each square. To do this click on the square tool and on **analyze**. Click on **measure** and start outlining each square of your image. The integrated relative density of each square will appear in the result window after clicking on **show results**. The first number is the area of your square and the second the average density (Mean). Record this value in a table opposite the size of the corresponding square.
7. Calculate OD × L and record the result in a table. Calculate the average length of the TRF using equation (7.1). The following is an example of a calculation using data from the table: $L_{TRF} = (2220 + 1800 + 1260)/180 + 200 + 210)$, i.e. $L_{TRF} = 5280/590$ and, therefore, $L_{TRF} = 8.9$ kb

Computer method

1. Align the X-ray film with the gel drawing made previously. The outline of the filter should be visible on the film. Transfer the positions of DNA standard bands onto the film.
2. Scan the autoradiogram with a computer scanner and save the image into your folder on the hard disk.
3. Open the program NIH Image. Open the **analyze** menu and select **option**. Select **user1**, **user2**, and **angle**. Select **max measurement** and set it to 2500.
4. Import your file into the program. Click on the **file** and **import** buttons. Select your file name and click on it. The image of your autograph with the labeled position of standards should appear on the desktop.
5. Load the Telometric macro. Go to **special** and click **load macros**. Browse the file system and find Telometric. Load it into the NIH Image program.
6. Click on **special** and start the **calibrate image** macro. Next click the mark on your autogram corresponding to the largest molecular weight standard. Enter the value of it (23 kb) in the window that has appeared and click OK. Click on the remaining standard molecular weight lines one at the time

and enter the corresponding molecular weight values in each window. These values are 9.4, 6.3, 4.3, and 2.3.

7. Press the **option/alt** key to enter these values into the program. You will see a window that states: calibration included five standards. Click the **OK** button.

8. Next you need to subtract the background gray scale. Outline the rectangular area on your image that contains a representative amount of background using **square tool**. Open **special** and click on the **specify background** macro. You should see a window that states: background region has been specified. Click the OK button.

9. Click the arrow on the NIH Image palette and outline the rectangular area that covers the first line of your sample. Adjust the size and the shape of the rectangle to include the entire area of the DNA "smear" in this line. Open **special** and click on the **outline first line** macro. An NIH Image window titled duplicate will appear with the first line outlined and numbered.

10. Click the arrow on the NIH Image and outline the next line as described in step 9. Open **special** and click on **outline next line**. Repeat this procedure for all lines of your autogram.

11. Open **special** and click on the **generate plots** macro. Plot windows that show graphs of a telomere length frequency distribution for each outlined line will appear. The y-axis of the graph is a relative telomere copy number and the x-axis is telomere length (in kilobases). Print this graph.

12. Open **special** and run the **generate statistic** macro. A table will appear that lists values (in kilobases) for the mean, median, mode, variance, and semi-interquartal for each line. Print this table. The mean value gives the mean telomere length. The mode value describes the most frequently occurring telomere length. The variance describes the variance in telomere length from which the error of the mean value can be calculated. The semi-interquartal and median values indicate the extent of heterogeneity of your telomeres. Large median and semi-interquartal values indicate large length heterogeneity of your telomeres.

13. Compare the values of the mean, median, mode, variance, and semi-interquartal with standard DNA and the DNA of your partner. Answer the following questions. Are your telomeres long or short? What is the molecular weight of the most frequent telomeres in your cells and the cells of your partner? Compare these values to the standard DNA. How homogenous are your telomeres and the telomeres of your partner and the telomeres of the standards given to you?

References

Blackburn, E.H. (1992) Telomerases. *Ann. Rev. Biochem.*, **61**, 113–29.
Brown, W.R.A. (1989) Molecular cloning of human telomeres in yeast. *Nature*, **338**, 774–6.

De Lange, T. and DePinho, R.A. (1999) Unlimited mileage from telomerase? *Science*, **203**, 947–9.

Evans, S.K. (2002) Telomeres. *Curr. Biol.*, **11**, R418.

Grant, J.D., Broccoli, D., Muquit, M., Manion, F.J., Tisdall, J., and Ochs, M.F. (2001) Telometric: a tool providing simplified, reproducible measurements of telomeric DNA from constant field agarose gels. *BioTechniques*, **31**, 1314–18.

Greider, C.W. and Blackburn, E.H. (1985) Identification of a specific telomere terminal transferase activity in *Tetrahymena* extracts. *Cell*, **43**, 405–13.

Griffith, J.D., Comeau, L., Rosenfield, S., Stansel, R.M., Bianchi, A., Moss, H. et al. (1999) Mammalian telomeres end in a large duplex loop. *Cell*, **97**, 503–14.

Levy, M.Z., Allsopp, R.C.B., and Futcher, A.B. (1992) Telomere end-replication problem and cell aging. *J. Mol. Biol.*, **225**, 951–60.

Linger, J. and Cech, T.R. (1998) Telomerase and chromosome end maintenance. *Curr. Opin. Genet. Devel.*, **8**, 226–32.

McEachern, M.J., Krauskopf, A., and Blackburn, E.H. (2000) Telomeres and their control. *Ann. Rev. Genet.*, **34**, 331–58.

Nugent, C.I. and Lundblad, V. (1998) The telomerase reverse transcriptase: components and replication. *Genes Devel.*, **12**, 1073–85.

Roche Molecular Biochemicals Co. (1999) *Instruction Manual for Telo TAGGG Telomere Length Assay*.

Surzycki, S.J. (2000) *Basic Techniques in Molecular Biology*. Springer-Verlag, Berlin, Heidelberg, and New York.

CHAPTER 8

RT-PCR of Human Genes

Introduction

The goal of this laboratory is to isolate and purify total human RNA and learn how to analyze gene expression. The source of RNA will be your cheek cells obtained from a saline mouthwash (a bloodless and non-invasive procedure). You will also learn how to work with RNA and determine its concentration and purity. The isolated RNA will be used in a RT-PCR (polymerase chain reaction) experiment for determining gene expression.

The exercise will be carried out in two laboratory periods. In the first period you will isolate and purify total RNA. The RNA will be analyzed using RNA gel electrophoresis for identifying major RNA species present in the cells. In the second laboratory period you will run an RT-PCR to determine expression of the β-actin gene in cheek cells. The product of this reaction will be analyzed by agarose gel electrophoresis.

Figure 8.1 presents an overall timetable for these experiments.

Background

Obtaining pure RNA is an essential step in the analysis of patterns of gene expression and understanding the mechanism of gene expression. Isolation of pure, intact RNA is one of the central techniques in today's molecular biology.

Two strategies of RNA isolation are usually employed: isolation of total RNA and isolation of mRNA. A typical eucaryotic cell contains approximately 10–20 pg of RNA, most of which is localized in the cytoplasm. Approximately 80–85 percent of eucaryotic RNA is ribosomal RNA, while 15–20 percent is composed of a variety of stable low molecular weight species such as transfer RNA and small nuclear RNA. Usually approximately 1–3 percent of the cell RNA is messenger RNA (mRNA) that is heterogeneous in size and base composition. Almost all eucaryotic mRNAs

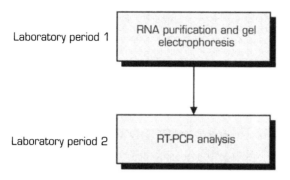

Figure 8.1 Schematic outline of the procedures used in the RT-PCR laboratory.

are monocystronic and contain a post-transcriptionally added polyadenylic acid (poly A) tract at their 3'-terminal. This poly-A 3'-tail permits separation and isolation of mRNA from all other classes of RNA present in the cell.

FIRST LABORATORY PERIOD

Experiment 1: purification of total RNA

Introduction

The goal of this experiment is to isolate total RNA from cheek cells. We will use a simple and quick procedure for obtaining a sufficient quantity of cellular RNA for analyzing its purity, carrying out RNA gel electrophoresis, and studying gene expression using an RT-PCR.

Background

Isolation of total RNA is most frequently used when pure RNA is required for experiments. This is because the techniques are less laborious and require less time to perform than isolation of mRNA. Various techniques for purification of total RNA are now in use and their application depends on the nature of the RNA required. For example, if RNA is going to be used for a quantitative RT-PCR, the intactness of the purified RNA is not critical, whereas intact RNA is required for cDNA library preparation or Northern blot analysis. Complete removal of DNA contamination is critical if RNA is used in an RT-PCR, but is not important in *in vitro* translation.

The physical and chemical properties of RNA and DNA are very similar. Thus, the basic procedures used in RNA purification are similar to those of DNA. All of the RNA purification methods incorporate the following steps.
1. Disruption of cells or tissue.
2. Effective denaturation of nucleoprotein complexes and removal of proteins.
3. Concentration of RNA molecules.
4. Determination of the purity and integrity of isolated RNA.

In addition, methods must include procedures that remove co-purified DNA from the preparation. In contrast to DNA purification, guarding against physical shearing of RNA molecules is not necessary because RNA molecules are much smaller and much more flexible than DNA molecules. In RNA protocols, strong physical forces during cell and tissue breakage are frequently used and the use of wide-mouth pipettes is not required. However, RNA isolation is much more difficult than DNA purification largely due to the sensitivity of RNA to degradation by internal and external ribonucleases. These enzymes are omnipresent and are very stable molecules that do not require any co-factors for their function. A crucial aspect of any procedure for RNA purification is fast and irreversible inactivation of endogenous RNases and protection against contamination with exogenous RNase during the isolation procedure. To these ends, all extraction buffers include powerful RNase inhibitors and all solutions and equipment used are treated to remove exogenous RNases.

Elimination of RNases

The most commonly used inhibitors included in extraction buffers for inhibiting endogenous RNase are as follows.

1. **Strong protein denaturation agents**. These include guanidinium hydrochloride and guanidinium isothiocyanate used at a concentration of 4M. These chaotropic agents can quickly inactivate endogenous RNases and contribute to denaturation of nucleoprotein complexes. In order to denature RNase irreversibly by these compounds, a high concentration of 2-mercapthoethanol is also included. This type inhibitor will be used in our RNA preparation procedure.

2. **Vanadyl–Ribonucleoside complexes**. Oxovanadium IV ions form complexes with any ribonucleoside and bind to most RNases, inhibiting their activity (Berger and Birkenmeier, 1979). A 10mM solution is used during cell breakage and is added to other buffers used in RNA isolation. Complexes can be used with deproteinizing agents (phenol or chloroform: isoamyl alcohol or CIA) and with chaotrophic agents. The compound is difficult to remove from purified RNA. Any residual amount of vanadyl will inhibit many enzymes used in subsequent RNA manipulations.

3. **Aurintricarboxylic acid (ATA)**. This compound binds selectively to RNase and inhibits its activity. ATA is usually incorporated into extraction buffers used for bacterial RNA preparations (Hallick et al., 1977). The inhibitor can affect certain enzymes and is not used if RNA will be used in primer extension or S1 nuclease experiments.

4. **Macaloid**. This is a naturally occurring clay (sodium magnesium lithofluorosilicate). Being negatively charged, it strongly absorbs all RNase. The macaloid and bound RNase are removed from the preparation by centrifugation (Marcus and Halvorson, 1967).

5. **Protein RNase inhibitors such as RNasin**. A protein originally isolated from human placenta, it inhibits RNase by non-competitive binding. It cannot be used in extraction buffers containing a strong denaturant. It is usually included in solutions used in the later stages of purification or in buffers used in storage or subsequent RNA manipulations. We will be using this inhibitor in the RT-PCR experiment.

The most frequent sources of exogenous RNase contamination are one's hands and bacteria and fungi present on airborne dust particles. The most frequently used inhibitors for removing exogenous RNase contamination are as follows.

1. **Diethyl pyrocarbonate (DEPC)**. DEPC causes enzyme inactivation by denaturing proteins. Inactivation of RNase is irreversible. The compound is used for removing RNase from solutions and glassware used in

RNA preparation. DEPC should be used with care because it is highly flammable and a suspected strong carcinogen.

2. RNaseZap™ or RNaseOff solutions. These commercially available reagents destroy RNases on contact very effectively. The decontamination solutions are not toxic and can be used for removing RNase from all surfaces and equipment. The compositions of these reagents are trade secrets.

Methods of RNA isolation

Three methods of RNA isolation or their modification are most frequently used: a guanidinium hot-phenol method, a high-salt lithium chloride method, and a TRI-Reagent™ method.

The guanidinium hot-phenol method is a modification of the procedure first described by Chirgwin et al. (1979) and Chomczynski and Sacchi (1987). This single-step extraction procedure takes advantage of the characteristic of RNA under acidic conditions to remain in the aqueous phase containing 4 M guanidine thiocyanate, while DNA and proteins are distributed into the phenol–chloroform organic phase. Distribution of DNA into the organic phase is particularly efficient if the DNA molecules are small. The method therefore uses a procedure for fragmenting DNA into molecules not larger than 10 kb. This method is used for isolating total RNA from a variety of procaryotic and eucaryotic cells. The efficiency of this method is very high (80–90 percent), affording purification of a large quantity of high-quality RNA.

The high-salt lithium chloride method is frequently used for isolating RNA from plant tissues that are particularly rich in various secondary products such as anthocyanins, phenolic compounds, polysaccharides, and latex. It has been shown that it is very difficult to isolate pure RNA from such plants using chaotropic agents (Schultz et al., 1994; Bugos et al., 1995). The procedure involves cell breakage in low pH, high salt buffer in the presence of RNase inhibitors. Protein and DNA are removed by acidic phenol–CIA extraction and RNA is recovered by lithium chloride precipitation.

The TRI-Reagent™ method is a single-step method of RNA isolation using a monophasic solution of phenol and guanidine isothiocyanate combined with precipitation of RNA by isopropanol in the presence of high salt (Chomczynski and Mackey, 1995). The method is particularly useful for fast isolation of RNA from numerous small samples and can be used with all types of cells and tissues. We will use a modification of this method (RNAwiz™).

Safety precautions

Each student should work only with his or her own cells. However, this restriction does not apply to purified RNA. Students that do not wish to work with their own cells can work with RNA isolated by anybody in the class.

RNAwiz™ contains phenol and guanidinium isothiocyanate. Both reagents are harmful to the skin. The reagents can be rapidly absorbed by and are highly corrosive to the skin. It initially produces a white softened area, followed by severe burns. Because of the local anesthetic properties of phenol, skin burns may not be felt until there has been serious damage. Gloves should be worn when working with this reagent. Because some brands of gloves are soluble or permeable to phenol, they should be tested before use. If TRI-Reagent™ is spilled on the skin flush off immediately with a large amount of water and treat with a 70 percent solution of PEG (polyethylene glycol) 4000 in water. Used reagent should be collected into a tightly closed glass receptacle and stored in a chemical hood until proper disposal.

Technical tips

The success of this experiment critically depends on rigorous control of RNase contamination. In order to prevent contamination of equipment and solutions with RNase, the following precautions should be taken.

1. Students should wear gloves at all time. Because gloves can be easily contaminated with RNase they should be changed frequently.
2. All tubes should be kept closed at all times.
3. Whenever possible disposable, certified RNase-free tubes, pipette tips, and plasticware should be used. Regular microfuge tubes and tips usually are not contaminated with RNase and they do not require special treatment if they are used from unopened bags.
4. All glassware should be treated with 0.1 percent DEPC water solution and autoclaved to remove DEPC. It is also possible to inactivate RNase by baking glassware at 180°C for at least 2 hours or overnight. Alternatively, RNase can be easily and efficiently eliminated from glassware, countertops, pipettors, and plastic surfaces using RNaseZap™ solution.
5. All solutions should be made with DEPC-treated water. Deionized water from a MilliQ RG apparatus can be used directly in all applications instead of DEPC-treated water because it does not contain RNase.
6. Since RNase treatment is frequently used in DNA isolation procedures, gel electrophoresis of DNA can cause electrophoresis tray and gel box contamination. Before their use for RNA gel electrophoresis, the gel tray and box should be decontaminated. To decontaminate the gel apparatus and gel-casting trays, treat them with RNaseZap™ solution. Instead of RNaseZap™ solution, the electrophoresis equipment can be treated with 0.2 N NaOH for 15 minutes and rinsed before use with RNase-free water.

Protocol

Collecting human cheek cells

1. Pour 13 ml of PBS into a 15 ml conical centrifuge tube. Transfer the solution into a paper cup. Pour all the solution into your mouth and swish vigorously for 30–40 seconds. Expel the PBS wash back into the paper cup.
2. Transfer the solution from the paper cup into a 25 ml Corex centrifuge tube and place it on ice.
3. Collect the cells by centrifugation at 6,000 r.p.m. for 10 minutes at 4°C.
4. Pour as much supernatant as possible back into the paper cup. Be careful not to disturb the cell pellet. Discard the supernatant from the paper cup into the sink. Invert the Corex centrifuge tube with cells on a paper towel to remove the remaining PBS.

RNA purification

1. Add 700 µl of RNAwiz^{TM}. Close the tube and mix by vortexing. Incubate for 5 minutes at room temperature. **Transfer the solution into a microfuge tube.**
2. Add 140 µl (0.2 volumes) of chloroform (not CIA) and mix by vortexing for 20 seconds. Incubate at room temperature for 10–15 minutes.
3. Centrifuge for 10 minutes at room temperature. After centrifugation, the mixture will be separated to two phases: a bottom phase containing chloroform and an upper aqueous phase containing RNA. **Note**: if an aqueous phase does not appear, add 100 µl of chloroform, vortex it for 20 seconds and repeat step 3.
4. Without disturbing the interphase, transfer the top aqueous phase to a fresh RNase-free microfuge tube. Add 350 µl (0.5 of starting volume) of RNase-free water and mix well by inverting the tube several times. Divide the solution into two new tubes. **Note**: if the combined volume of aqueous phase and water is less then 800 µl it is not necessary to divide this mixture into two tubes. Instead add 700 µl of isopropanol to the tube, mix well, and let it stand at room temperature for 10 minutes. Next, follow the procedure from step 10
5. Precipitate RNA by the addition of 350 µl of isopropanol to each tube. Mix by inverting the tubes several times and incubate at room temperature for 10 minutes.
6. Place the tubes into the centrifuge, orienting the attached end of the tube lid away from the center of rotation. Centrifuge for 15 minutes at room temperature to pellet the RNA.
7. Remove the tubes from the centrifuge. Remove the supernatant using a P200 Pipetman. Wash the pellet with 700 µl of cold 70 percent ethanol. Add ethanol to each tubes and mix by inverting several times.

8. Place the tubes into a microfuge and centrifuge for 5 minutes at room temperature. Remove ethanol with a P200 Pipetman.

9. Place the tubes into the centrifuge, making sure that the side containing the pellet faces away from the center of rotation. Start the centrifuge until it reaches 500 r.p.m. (1–2 seconds). This will collect ethanol from the sides of the tube. Remove ethanol using a P200 Pipetman equipped with capillary tip.

10. Prewarm RNAsecure solution in a 60°C water bath for 5 minutes. Add 15 µl of this solution to one tube and dissolve the pelleted RNA. Transfer the solution to the second tube and dissolve the pellet.

11. As fast as possible heat the sample to 60°C for 10 minutes to inactivate potential RNase contamination. Store the RNA sample at −70°C. **Note**: since RNAsecure only inactivates RNases at 60°C the best results are obtained using prewarmed solution and transferring the tube as quickly as possible to a 60°C water bath.

12. Determine the concentration of RNA by measuring the absorbance at 260 nm. Initially use a 1:100 dilution of the sample in PBS. The absorbance reading should be in the range 0.1–1.5. Calculate the concentration of RNA using the equation $N = A_{260}/\varepsilon_{260}$, where N is the RNA concentration in micrograms per milliliter, ε_{260} is the RNA extinction coefficient, and A_{260} is the absorbance reading (corrected for dilution). The absorption coefficient for total RNA is usually taken to be $0.025\,\mu g^{-1}\,cm^{-1}$ giving a solution of $40\,\mu g\,ml^{-1}$ of RNA an absorbance of 1.0 (e.g. $1/0.025 = 40\,\mu g\,ml^{-1}$).

13. To determine the purity of the RNA, measure the absorbance at 260 nm, 280 nm, and 234 nm and calculate the 260 nm:280 nm and 260 nm:234 nm ratios. The concentration of your RNA should be between 0.1 and $0.3\,\mu g\,\mu l^{-1}$.

References

Berger, S.L. and Birkenmeier, C.S. (1979) Inhibition of intractable nucleases with ribonucleoside–vanadyl complexes: isolation of messenger ribonucleic acid from resting lymphocytes. *Biochemistry*, **18**, 5143–9.

Bugos, R.C., Chiang, V.L., Zhang, X.-H., Campbell, E.R., Podilla, G.K., and Campbell, W.H. (1995) RNA isolation from plant tissues recalcitrant to extraction in guanidine. *BioTechniques*, **19**, 734–7.

Chirgwin, J., Przybylska, A., MacDonald, R., and Rutter, W. (1979) Isolation of biologically active ribonucleic acid from sources enriched in ribonuclease. *Biochemistry*, **18**, 5294–9.

Chomczynski, P. and Mackey, K. (1995) Modification of the TRI Reagent™ procedure for isolation of RNA from polysaccharide- and proteoglycan-rich sources. *BioTechniques*, **19**, 942–5.

Chomczynski, P. and Sacchi, N. (1987) Single-step method of RNA isolation by acid guanidinium thiocyanate–phenol–chloroform extraction. *Anal. Biochem.*, **162**, 154–9.

Hallick, R.B., Chelm, B.K., Gray, P.W., and Orozco, E.M. (1977) Use of aurintricarboxylic acid as an inhibitor of nucleases during nucleic acid isolation. *Nucleic Acids Res.*, **4**, 3055–64.

Marcus, L. and Halvorson, H.O. (1967) Resolution and isolation of yeast polysomes. In *Methods in Enzymology*, Vol. 12A, L. Grossman and K. Moldave (eds), pp. 498–502. Academic Press, New York and London.

Schultz, D.J., Craig, R., Cox-Foster, D.L., Mumma, R.O., and Medford, J.I. (1994) RNA isolation from recalcitrant plant tissue. *Plant Mol. Biol. Rep.*, **12**, 310–16.

Experiment 2: RNA agarose gel electrophoresis

Introduction

In this experiment you will continue the analysis of total RNA using agarose gel electrophoresis. We will use a native agarose gel system that is easy to prepare and does not use toxic chemicals. You will run a mini-gel that is sufficient for demonstrating the presence of stable species of RNA in your preparation. Native gels can also be used as an analytical tool for assessing the efficiency of RNA purification.

Background

Gel electrophoresis of RNA molecules requires techniques that are different from those used for DNA. In order to separate RNA molecules according to their size it is necessary to maintain their complete denaturation before and during electrophoresis. Non-denatured RNA can form secondary structures such as "hairpins" that profoundly influence their electrophoretic mobility. A number of denaturants have been used. Among these are glyoxal with DMSO (McMaster and Carmicheal, 1977), formaldehyde (Lehrbach et al., 1977; Rave et al., 1979), and methylmercuric hydroxide (Bailey and Davidson, 1977; Thomas, 1980). Formaldehyde and glyoxal-DMSO are presently used more often than the highly toxic methylmercuric hydroxide.

The buffers used for RNA electrophoresis differ from those used for DNA. These buffers are of very low ionic strength, frequently resulting in the creation of a pH gradient along the length of the gel that causes overheating of the gel and distortion of RNA bands. To prevent this, RNA gels are usually run at low field strength ($<5\,V\,cm^{-1}$) using a large volume of buffer and constant stirring to prevent gradient formation.

It is also possible to run native RNA agarose gels that do not include toxic denaturants in the agarose gel. The native gel system is simple and, in general, does not affect electrophoretic separation of RNA (Liu and Chou, 1990).

The glyoxal-DMSO method does not use toxic chemicals, but it is more difficult to use than the native method. This method requires very careful control of pH during electrophoresis to a pH below 8. This is because glyoxal denatures RNA by binding covalently to the guanine residue, forming products that are stable only at a pH below 8. At a pH above 8, glyoxal dissociates from RNA. Submarine gels require continuous recirculation and mixing of electrophoresis buffer in order to maintain the pH within an acceptable limit. In addition, commercially available glyoxal must be purified before use in order to remove glyoxylic acid, which is readily formed by oxidation and degrades RNA. The electrophoresis time is longer than for native gels.

Safety precautions

Agarose gel contains ethidium bromide, which is a mutagen and suspected carcinogen. Contact with the skin should be avoided. Students should wear gloves when handling ethidium bromide solution and gels containing ethidium bromide. Discard the used gel into the designated container.

For safety purposes, the electrophoresis apparatus should always be placed on the laboratory bench with the positive electrode (red) facing away from the investigator, that is away from the edge of the bench. To avoid electric shock always disconnect the red (positive) lead first.

Ultraviolet (UV) light can damage the retina of the eye and cause severe sunburn. Always use safety glasses and a protective face shield to view the gel. Work in gloves and wear a long-sleeved shirt or laboratory coat when operating UV illuminators.

Technical tips

The most common problems with gel electrophoresis of RNA are inadequate denaturation of the samples and overloading the gel with RNA. Inadequate denaturation will appear either as multiple rRNA bands or rRNA bands that appear to be smeared but at a correct ratio. Overloading the gel will result in very broad rRNA bands that run on the gel with excessive "smearing." Bands could have a U-shape appearance and their mobility might be faster than expected from their base number.

Sample degradation will be indicated by an incorrect ratio between 28S and 18S rRNA or, in more severe cases, the total disappearance of these bands. Students can use even a severely degraded sample for the RT-PCR experiment.

Figure 8.2 presents an image of native gel electrophoresis of total human RNA isolated from cheek cells.

Protocol

1. Prepare a mini-gel using a casting tray not larger than 7.5 cm × 7.5 cm and a thin gel (0.2 cm). Seal the ends of the gel-casting tray with tape.

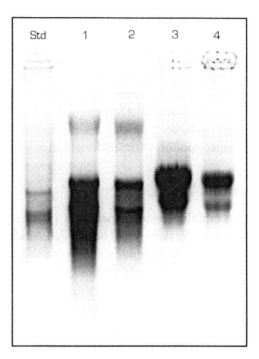

Figure 8.2 Native agarose gel electrophoresis of human total RNA. Lane 1 standard RNA and lanes 2–4 RNA isolated from cheek cells. Electrophoresis was carried out on a 1 percent agarose gel for 1 hour at 5 V cm^{-1}.

Regular labeling tape or electrical insulation tape can be used. Use a mini-gel comb with wells 0.2–0.5 cm long and 1 mm (or less) wide. Check that the bottom of the comb is approximately 0.5 mm above the gel bottom. To adjust this height it is most convenient to place a plastic charge card (for example MasterCard) at the bottom of the tray and adjust the comb height to a position where it is easy to remove the card from under the comb. Wipe the comb with RNaseZap^{TM} immediately before use.

2. Prepare 500 ml of one times TBE (Tris–borate EDTA) buffer by adding 50 ml of a ten times TBE stock solution to 490 ml of RNase-free water.

3. Prepare a 1 percent agarose gel. Place 15 ml of the buffer into a 100 ml flask and add 150 mg of agarose powder. Melt the agarose by heating the solution in a microwave oven at full power for 1–2 minutes until the agarose is fully dissolved. If evaporation or spillage occurs during melting, adjust the volume to 15 ml with deionized water.

4. Cool the agarose solution to approximately 60°C and add 1 µl of ethidium bromide stock solution. Slowly pour the agarose into the casting tray. Remove any air bubbles by trapping them in a 10 ml pipette. Place the comb 1 cm away from one end of the gel. Allow the gel to solidify for 20–30 minutes. **Note**: native agarose gels should be as thin as possible (2–3 mm) in order to shorten the electrophoresis time.

5. Remove the tape from the ends of the gel-casting tray and place the tray on the central supporting platform of the gel box. Add electrophoresis buffer until it reaches a level approximately 2–3 mm above the surface of the gel.
6. Prepare the RNA sample for electrophoresis in a 0.2 ml thin-wall, RNase-free tube. Prepare the sample as follows. Add 2 μl of five times RNA sample buffer to a sterile microfuge tube and 1 or 2 μl of your RNA and fill it up with sterile RNase-free water to a total volume of 10 μl.
7. Place the tube into a LightCycler and incubate at 65°C for 10 minutes. Transfer it immediately to ice. Incubate on ice for at least 2 minutes. Centrifuge the tube for 20 seconds to collect condensation and place the tube back on ice until ready to load onto the gel.
8. Load the samples into the wells using a yellow, RNase-free tip. Place the tip **under** the surface of the electrophoresis buffer and **above** the well. Expel the sample slowly, allowing it to sink to the bottom of the well. Take care not to spill the sample into a neighboring well. During sample loading, it is very important not to place the end of the tip into the sample well or touch the edge of the well with the tip. This can damage the well, resulting in uneven or smeared bands.
9. Place the lid on the gel box and connect the electrodes. RNA will travel towards the positive (red) electrode. Turn on the power supply. Adjust the voltage to approximately $5\,V\,cm^{-1}$. For example, if the distance between electrodes (not the gel length) is 20 cm, in order to obtain a field strength of $5\,V\,cm^{-1}$ the voltage should be set to 100 V. Continue electrophoresis until bromophenol blue moves at least two-thirds of the length of the gel. It will take the tracking dye approximately 30 minutes to reach this position.
10. Turn the power supply off and first disconnect the positive (red) and then the negative lead from the power supply. This order of disconnecting leads prevents the occurrence of accidental electrical shock. Remove the gel from the electrophoresis chamber. You can photograph the gel using a computer imaging system to record the results.
11. Two sharp bands will appear on the gel, 28S RNA (4.7 kb) and 18S RNA (1.9 kb). The 5S and 5.8S RNA bands are located on the leading edge of the gel, running together with tRNA. **Note**: the integrity of the prepared RNA samples and lack of RNA degradation can be easily judged from the appearance of rRNA bands. Degradation of the sample appears as diffused 28S and 18S bands or an incorrect ratio of stain between rRNA bands. This ratio should be approximately 2 : 1 for 28S and 18S, respectively (see Fig. 8.2).

References

Bailey, J.M. and Davidson, N. (1977) Methylmercury as reversible denaturating agent for agarose gel electrophoresis. *Anal. Biochem.*, **70**, 75–85.

Liu, Y.-C. and Chou, Y.-C. (1990) Formaldehyde in formaldehyde/agarose gel may be eliminated without affecting the electrophoretic separation of RNA molecules. *BioTechniques*, **9**, 558–60.

Lehrbach, H., Diamond, D., Wozney, J.M., and Boedtker, H. (1977) RNA molecular weight determination by gel electrophoresis under denaturing conditions, a critical reexamination. *Biochemistry*, **16**, 97–101.

McMaster, G.K. and Carmicheal, G.C. (1977) Analysis of single- and double-stranded nucleic acids on polyacrylamide and agarose gels and acridin orange. *Proc. Natl Acad. Sci. USA*, **74**, 4835–8.

Rave, N., Crkvenjakov, R., and Boedtker, H. (1979) Identification of procollagen mRNAs transferred to diazobenzyloxymethyl paper from formaldehyde gels. *Nucleic Acids Res.*, **6**, 3559–67.

Thomas, P.S. (1980) Hybridization of denatured RNA and small DNA fragments transferred to nitrocellulose. *Proc. Natl Acad. Sci. USA*, **77**, 5201–5.

SECOND LABORATORY PERIOD

Experiment 3: running an RT-PCR

Introduction

During this laboratory period you will perform an RT-PCR using the RNA purified in experiment 1. You will analyze β-actin genes expression in cheek cells. There are six known actin proteins in mammalian cells: two sarcomeric muscle actins (α-skeletal and α-cardiac), two smooth muscle actins (α and γ), and two non-muscle cytoskeletal actins (β and γ). Three genes have been mapped for human actin protein: β-actin gene on chromosome 7, α-skeletal actin gene on chromosome 1, and α-cardiac antigen on chromosome 15. The β-actin gene on chromosome 7 was mapped to the 7p15 position. In addition to one functioning β-actin gene, there are approximately 20 pseudo-genes widely distributed in the human genome. Only four β-actin pseudo-genes were mapped to other chromosomes (chromosomes 5, 13, and 18).

We will be using the "One Tube RT-PCR Kit" from Roche Molecular Biochemicals Co. This kit is designed for sensitive, quick, and reproducible analysis of RNA with high fidelity. The one step reaction system uses avian myoblastosis virus (AMV) transcriptase for first-strand synthesis and Expand™ High Fidelity enzyme blend, which consists of *Taq* DNA polymerase and *Pwo* DNA polymerase, for the PCR.

The expected length of the product of our RT-PCR reaction should be 557 bases (see Figure 8.3 for details). Products of different size can also appear. This will depend on the structure of your gene (a missing processing site at the intron–exon boundary) or partial expression of pseudo-genes (mRNA synthesis).

Background

Sensitive methods for the detection and analysis of RNA molecules are an important aspect of most cell/molecular biology studies. Commonly used methods include *in situ* hybridization, Northern blots, dot or slot blot analysis, S1 nuclease analysis, and RNase protection assays. *In situ* hybridization is very sensitive, but is a rather difficult technique. Other common methods lack sensitivity for detecting small amounts of RNA.

The adaptation of PCR methodology to the investigation of RNA provides a method having speed, efficiency, specificity, and sensitivity. Since RNA cannot serve as a template for a PCR, reverse transcription is combined with PCR to make RNA into a complementary DNA (cDNA) suitable for PCR. The combination of both techniques is named RT-PCR.

The process of RT-PCR has proven to be invaluable for detecting gene expression, for amplifying RNA sequences prior to subcloning and analysis, and for the diagnosis of infectious agents or genetic diseases. This technique is the most sensitive procedure for determining the presence or absence of RNA templates or quantifying the level of gene expression. Furthermore, RT-PCR allows cloning of rare messages without having to construct cDNA libraries. The use of RT-PCR for analyzing genetic diseases is particularly advantageous because it circumvents inefficient amplification of long DNA fragments caused by long introns and provides additional information about phenomena such as alternative splicing.

Two different techniques are used for RT-PCR.

1. **Two step RT-PCR**. The synthesis of cDNA is performed with reverse transcriptase from AMV, moloney murine leukemia virus (M-MuLV), or *Thermus thermophilus* (*Tth*) DNA polymerase in the first step, followed by PCR with an appropriate thermostable DNA polymerase. The two-step reaction requires that the reaction tube is opened after cDNA synthesis and reagents are added for the PCR part of the procedure. This is inconvenient and increases the risk of contamination.

2. **One-step RT-PCR**. The cDNA synthesis and the PCR are performed together in a single tube. Two techniques are in use for running one-step RT-PCRs. The first technique uses *Tth* DNA polymerase for carrying reverse transcription and PCR reactions. The second uses AMV reverse transcriptase and *Taq* DNA polymerase.

The first method relies on the ability of the *Tth* DNA polymerase to use an RNA template for DNA synthesis (reverse transcription) as well as to use a DNA template (PCR). This enzyme is thermostable allowing both reactions to be carried at elevated temperature. Carrying out the reaction at elevated temperature helps unravel secondary structures of RNAs, thereby allowing the synthesis of longer products and increasing the efficiency synthesis of the first strand of DNA. However, low processivity of the *Tth* DNA polymerase limits the RT-PCR products to less than 1.0 kb.

The second method uses AMV reverse transcriptase and *Taq* DNA polymerase. This technique allows amplification of fragments of up to 2.0 kb with lower error rates as compared to the use of *Tth* DNA polymerase. However, the reverse transcription step of the reaction has to be performed at 42°C and, therefore, is strongly affected by the secondary structure of mRNA.

Reverse transcriptases

Reverse transcriptases are RNA-dependent DNA polymerases that have predominantly been used for catalyzing first-strand cDNA (complementary DNA) synthesis. However, reverse transcriptases are also capable of synthesizing a DNA strand complementary to a primed single-stranded DNA.

Three different enzymes with reverse transcriptase activity are now

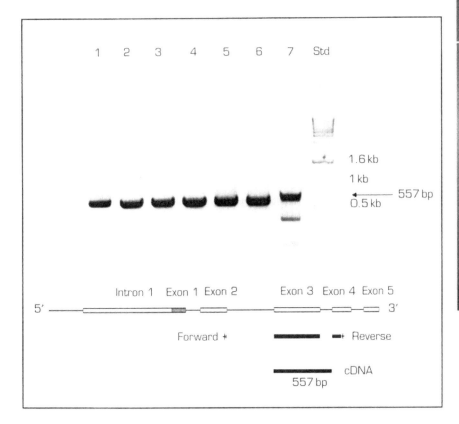

Figure 8.3 Agarose gel electrophoresis of the products of RT-PCR reactions. The structure of genomic DNA encoding the human gene for β-actin is shown. The size of the expected cDNA and the position of primers in the genomic DNA are indicated. The expected product of PCR amplification, when genomic DNA is used, is 1,200 bp.

commercially available: the viral reverse transcriptases (RTases) from avian myoblastosis virus (AMV), M-MuLV, and the heat-stable DNA polymerase derived from *T. thermophilus*. All these enzymes require different pH, salt concentration, and incubation temperatures for optimal activity.

The AMV and M-MuLV viral RTases are highly processive and are able to synthesize cDNAs of up to 10 kb. *Tth* DNA polymerase is able to synthesize cDNA in the range of 1.0–2.0 kb, which is sufficient since fragments of <1 kb are usually used for PCRs.

The unique advantage of the *Tth* DNA polymerase is its ability to perform both reverse transcription and PCR amplification in a one-step reaction.

Priming of a reverse transcriptase reaction

There are three types of primers that may be used for reverse transcription.
1. **Oligo(dT)12–18 primer.** This primer binds to the endogenous poly

(A)+ tail at the 3'-end of mammalian mRNA. A reaction with this primer frequently produces a full-length cDNA product.

2. **Random hexanucleotide primers**. These primers can bind to mRNA templates at any complementary site and will give partial length (short) cDNAs. These primers may be better for overcoming the difficulties caused by template secondary structure. The random primers may also transcribe more 5'-regions of the RNA.

3. **Specific oligonucleotide primers**. These primers can be used for selectively priming the RNA of interest. This approach has been used very successfully in diagnostic assays, as well as in basic research. We will be using specific primers for human β-actin genes.

Safety precautions

The agarose gel contains ethidium bromide, which is a mutagen and suspected carcinogen. Contact with the skin should be avoided. Students should wear gloves when handling ethidium bromide solution and gels containing ethidium bromide. Discard the used gel into the designated container.

For safety purposes, the electrophoresis apparatus should always be placed on the laboratory bench with the positive electrode (red) facing away from the investigator, that is away from the edge of the bench. To avoid electric shock always disconnect the red (positive) lead first.

UV light can damage the retina of the eye and cause severe sunburn. Always use safety glasses and a protective face shield to view the gel. Work in gloves and wear a long-sleeved shirt or laboratory coat when operating UV illuminators.

Technical tips

The Titan one-tube RT-PCR Kit from Roche Molecular Biochemicals Co. is used in this experiment. The reverse transcriptase reaction is performed using AMV polymerase at elevated temperatures minimizing the influence of the secondary structure of mRNA on the synthesis of the first strand. The Expand^{TM} High Fidelity enzyme blend is used for the PCR part of the reaction. Synthesis of cDNA is much faster than in the two-step system and also faster than synthesis achieved by many other one-step kits. This makes it an ideal kit for use in a class environment when performing RT-PCR reactions and product analysis during the same laboratory period.

RT-PCR reactions should when possible make use of primers that do not amplify genomic DNA or produce fragments that are larger than fragments amplified from cDNA. This is possible if primers are positioned in different exons with very large introns located between them. This is necessary because it is very difficult to remove genomic DNA from purified RNA to the

Table 8.1 Master mix 1 (MX1)

Ingredient concentration	For one reaction	For three reactions	Final concentration
4dNTP (ten times)	2.0 µl	6.0 µl	200.0 mM
DTT	1.0 µl	3.0 µl	5.0 mM
RNase inhibitor	0.5 µl	1.5 µl	5.0 units
Primers mix	1.0 µl	3.0 µl	20.0 µM
RNA	up to 3.0 µl	–	At least 150.0 ng
Water	5.0 µl	15.0 µl	
Total	12.5 µl	28.5.0 µl	

Table 8.2 Master mix 2 (MX2)

Ingredient concentration	For one reaction	For three reactions	Final concentration
Five times RT-PCR buffer	5.0 µl	15.0 µl	One times
Enzyme mix	0.5 µl	1.5 µl	One times
Water	7.0 µl	21.0 µl	
Total	12.5 µl	37.5 µl	

extent that it will not be amplified by a PCR. This principle is very well illustrated in the experiment described. The amplification of residual genomic DNA, when present in RNA, should result in a 1,200 bp DNA fragment, while amplification of cDNA results in a 557 bp fragment (see Figure 8.3). However, the presence of a larger than expected fragment is not always the result of the amplification of contaminating genomic DNA. These fragments can result from amplification cDNA obtained from unprocessed RNA or RNA originating from pseudo-genes.

Protocol

1. Label two RNase-free 1.5 ml microfuge tubes MX1 and MX2 (master mix 1 and master mix 2, respectively). Place the tubes into an ice bucket.
2. Prepare master mix 1 and master mix 2 as described in Tables 8.1 and 8.2. Remember to start assembling the reactions by the addition of water and buffer. Add the enzyme (for MX2) last. Mix all of the ingredients by pipetting up and down. Centrifuge the tubes for 30 seconds.
3. You will need to run your RNA in a single reaction. Therefore, you will have two reactions per group. Prepare three reactions per group in order to compensate for pipetting errors.
4. Label two RNase-free thin-walled tubes with your group number and

Table 8.3 RT-PCRs

Ingredient cencentration	Your reaction	Partner's reaction	Final concentration
MX1 mix	9.5 µl	9.5 µl	One times
MX2 mix	12.5 µl	12.5 µl	One times
Your RNA	3.0 µl	–	100–200 ng
Partner's RNA	–	3.0 µl	100–200 ng
Total	25.0 µl	25.0 µl	

place them on ice. Assemble RT-PCR reactions in these tubes using MX1 and MX2 as indicated in Table 8.3. Mix by pipetting up and down and centrifuge briefly.

5. Place the tubes into the thermocycler and run the RT-PCR reaction using the following program: 50°C for 30 minutes and 94°C for 30 seconds. Cycle 1 with denaturing at 94°C for 30 seconds, annealing at 50°C for 30 seconds, and elongation at 68°C for 1 minute ($n = 10$). Cycle 2 with denaturing at 94°C for 30 seconds, annealing at 50°C for 30 seconds, and elongation at 68°C for 2.5 minutes ($n = 25$) and end with 5 minutes at 68°C. Store tubes in a –70°C freezer.

Analysis of the results of the RT-PCR

1. Prepare a mini-gel using a casting tray not larger than 7.5 cm × 7.5 cm. Seal the ends of the gel-casting tray with tape. Regular labeling tape or electrical insulation tape can be used. Check that the bottom of the comb is approximately 0.5 mm above the gel bottom. To adjust this height, it is most convenient to place a plastic charge card (for example MasterCard) at the bottom of the tray and adjust the comb height to a position where it is easy to remove the card from under the comb.

2. Prepare a 1.6 percent agarose mini-gel in one times TAE (Tris–acetate EDTA) buffer. Use 30 ml of agarose solution. Weigh 480 mg of agarose and add it to 30 ml of one times TAE buffer. Dissolve agarose in a microwave oven, adjust the volume to 30 ml with water, and add 1 µl of ethidium bromide. Pour the agarose into the casting tray. Allow the gel to solidify for 20–30 minutes.

3. Add 5 µl of stop solution to each tube with PCR reactions. Mix by pipetting up and down. Load prepared samples onto the gel. If only two or three samples are loaded onto the gel, use wells in the center of the gel.

4. Load 6 µl of a size standard (1 kb ladder) into the well to the left of your samples. Run gel electrophoresis for 20–30 minutes at 60–80 V. Photograph the gel and analyze the results.

5. A typical result of an RT-PCR gel is presented in Fig. 8.3. Line 7 contains a RT-PCR using mRNA of human β-actin.

Appendix

This appendix provides a list of the equipment and supplies necessary for running each laboratory. Because many of the experiments described use the same equipment and supplies these are listed only once. Recipes for the solutions are also given for each exercise. Each recipe is listed only once for the laboratory when it is first required.

DNA Purification

Equipment and supplies

1. 8HQ (8-hydroxyquinoline) free base (Sigma Co., no. H6878). Do not use hemisulfate salt of the 8HQ.
2. Sterilized 15 ml and 50 ml conical polypropylene centrifuge tubes (e.g. Corning, nos 25319-15 and 25330-50).
3. Corex 25 ml centrifuge tubes with Teflon-lined caps (Corex, no. 8446-25).
4. EDTA (ethylenediaminetetra-acetic acid) (0.5 M at pH 8.0) (Ambion Inc., no. 9261).
5. Glass hooks. Glass hooks are made from Pasteur pipettes in the following way. First, place the end of a pipette horizontally into a Bunsen flame and seal it. Next, holding the pipette at a 45° angle, insert 0.5 cm of the tip into the flame. The end of the pipette will slowly drop under gravity forming a hook.
6. Ten times PBS (pH 7.4) (Ambion, no. 9625). The pH of the PBS solution is critical for collecting cheek cells. Preparation of this solution from basic ingredients is not recommended.
7. Redistilled, water-saturated phenol (Ambion Inc., no. 9712). Water-saturated phenol is preferable to the crystalline form because it is easier and safer to prepare buffered phenol from it. Water-saturated phenol can be stored indefinitely in a tightly closed, dark bottle at −70°C.
8. Proteinase K (Ambion Inc., no. 2546). Ambion Inc. is a low-cost source

of proteinase K. The enzyme is supplied in storage buffer containing 50 percent glycerol at a concentration of 20 mg ml^{-1}. Proteinase K solution should be stored at –20°C. Proteinase K remains active for several years at –20°C.

9. DNase-free ribonuclease A (Sigma, no. R4642). RNase is supplied in storage buffer with 50 percent glycerol at a concentration of 10 mg ml^{-1}. The enzyme can be stored indefinitely in a –20°C freezer.

10. DNase-free ribonuclease T1 (Ambion Inc., no. 2280). The enzyme is supplied in a storage buffer with 50 percent glycerol at concentration of 2000 units ml^{-1} and can be stored indefinitely in a –20°C freezer.

11. Sarcosyl (*N*-lauroylsarcosine) sodium salt (Sigma, no. L 5125).

Solutions to prepare

1. CIA (chloroform:isoamyl alcohol). Mix 24 volumes of chloroform with 1 volume of isoamyl alcohol. Because chloroform is light sensitive and very volatile, the CIA solution should be stored in a brown glass bottle, preferably in a fume hood.

2. Dilution buffer: 20 mM Tris–HCl (pH 8.5) and 100 mM Na$_2$ EDTA (pH 8.0). Prepare as described for lysis buffer.

3. Lysis buffer: 20 mM Tris (pH 8.5), 100 mM Na$_2$ EDTA (pH 8.0), 120 mM NaCl, and 1.2 percent Sarcosyl. Add the appropriate amount of 1 M Tris–HCl stock solution and 0.5 M EDTA stock to the water. Check the pH of the lysis buffer and titrate it to pH 8.5 with concentrated NaOH if necessary. Add NaCl and sterilize by autoclaving for 20 minutes. Do not add Sarcosyl to the stock solution. It will be added after resuspension of the cells. Store at 4°C.

4. Phenol 8HQ. Water-saturated, twice-distilled phenol is equilibrated with an equal volume of 0.1 M sodium borate. Sodium borate should be used rather than the customary 0.1 M Tris solution because of its superior buffering capacity at pH 8.5, its low cost, its antioxidant properties, and its ability to remove oxidation products during the equilibration procedure. Mix an equal volume of a water-saturated phenol with 0.1 M sodium borate in a separation funnel. Shake until the solution turns milky. Wait for the phases to separate and then collect the bottom phenol phase. Add 8HQ to the phenol at a final concentration 0.1 percent (v/w). Phenol 8HQ can be stored in a dark bottle at 4°C for several weeks. Store at –70°C for long-term storage. The solution can be stored for several years at –70°C.

5. Twenty percent (w/v) Sarcosyl stock. Dissolve 40 g of Sarcosyl in 100 ml of double-distilled or deionized water. Adjust to 200 ml with double-distilled or deionized water. Sterilize by filtration through a 0.22 μm filter. Store at room temperature.

6. Sodium acetate (3 M). Sterilize the solution by autoclaving and store at 4°C.

7. TE buffer: 10 mM Tris–HCl (pH 7.5 or 8.0) and 1 mM Na_2 EDTA (pH 8.0). Sterilize by autoclaving and store at 4°C.

DNA Fingerprinting: Multi-locus Analysis

Equipment and supplies

1. Agarose powder SeaKem™ LE (FMC BioProducts, no. 50001) or equivalent.
2. Anti-DIG (digoxigenin) alkaline phosphatase (750 u ml^{-1}) (Roche Molecular Biochemicals, no. 1093 274). The stock solution should not be frozen. Store at 4°C.
3. Capillary tubes (25 µl) (Idaho Technology, no. 1709).
4. CDP-Star solution (Tropix Co., no. MS100R). Store the solution in the dark at 4°C. CDP-Star is easily destroyed by ubiquitous alkaline phosphatase. Wear gloves and use sterilized tips when handling CDP-Star solution.
5. Dig Easy Hyb solution (Roche Molecular Biochemicals, no. 1603 558).
6. Dig Wash and Block Buffer Set (Roche Molecular Biochemicals, no. 1585 762). It is possible to prepare each reagent from basic ingredients, but the cost of it will be higher than the cost of the kit.
7. Ten times DIG-dUTP labeling mixture (Roche Molecular Biochemicals, no. 1 227 065).
8. DIG-labeled control DNA (Roche Molecular Bichemicals, no. 1093 657).
9. DNA Ladder (1 kb) (Life Technologies, no. 15615-016).
10. Ten times dNTP mix for an air cycler (Idaho Technology, no. 1774).
11. Ethidium bromide (Sigma Co., no. E 8751).
12. Ficoll 400 (Sigma Co., no. F 4375).
13. Gel electrophoresis apparatus (minimum 13 cm × 20 cm gel size) with power supply (e.g. Owl Scientific, no. A1). One per group.
14. *Hae*III restriction enzyme (NEB, no. 108S or equivalent).
15. Male and female human genomic DNA (Sigma, nos D-3160, D-3035, or equivalent).
16. Hybridization bottles (150 mm × 35 mm) (e.g. HyBaid Co., no. H9084 or equivalent).
17. A hybridization oven (e.g. HyBaid Co., no. H9320 or equivalent). The oven should be capable of rotation at variable speeds.
18. M13mp RF1 DNA (Amersham Pharmacia Biotech, no. 27-1547-01).
19. MagnaGraph nylon membrane (0.22 µ pore size) (Osmonics/MSI., no. NJTHY00010).
20. NBT solution (Roche Molecular Biochemicals, no. 1 383 213).
21. Ten times PCR (polymerase chain reaction) buffer mix for the air cycler (Idaho Technology, no. 1781).

22. A plastic bag sealer (Fisher Scientific, no. 01-812-13 or equivalent).
23. Plastic bags (e.g. Kapak Co., no. 402 or Roche Molecular Biochemicals, no. 1666 649).
24. Primers: M13 V F, GGTACATGGGTTCCTATT and M13 V R, CCCTTATTAGCGTTTGCCAT.
25. Pyrex glass dishes (two per group).
26. Rotary platform shakers.
27. A Stratalinker® ultraviolet (UV) oven (Stratagen Co., no. 400071 or equivalent). In order to cross-link DNA to the nylon membrane efficiently the UV source should be capable of delivering $120\,mJ\,cm^{-2}$. Excessive cross-linking will decrease the hybridization efficiency.
28. Taq DNA polymerase.
29. Whatman 3MM chromatography paper (Whatman Co., no. 3030917).
30. X-phosphate solution (Roche Molecular Biochemicals, no. 1 383 221).
31. X-ray film BioMax Light (Kodak, 8 × 10 in no. 178–8207 or equivalent).

Solutions to prepare

1. Buffer A. Add 100 ml of buffer 1 (Roche Molecular Biochemicals Set) to 900 ml of sterilized distilled water. Store at 4°C.
2. Buffer B (blocking solution). Add 10 ml of blocking solution (Dig Wash and Block Buffer Set, Roche Molecular Biochemicals) to 80 ml of sterilized distilled water. Add 10 ml of maleic acid buffer (bottle 2, Dig Wash and Block Buffer Set, Roche Molecular Biochemicals). Always prepare freshly.
3. Buffer C (detection buffer): 0.1 M Tris–HCl (pH 9.5) and 0.1 M NaCl.
4. Denaturation solution: 0.5 N NaOH and 1.5 M NaCl. Prepare the solution using 10 N NaOH. The solution can be stored at room temperature for a few months. If a white precipitate forms, the solution should be discarded.
5. Depurination solution: 0.5 N HCl.
6. Ethidium bromide stock. ($5\,mg\,ml^{-1}$): 100 mg ethidium bromide and 20 ml water. Dissolve the powder in the water by stirring under a chemical hood. Store at room temperature in a tightly closed, dark bottle.
7. Neutralization solution: 0.5 M Trisma base and 1.5 M NaCl. Add 60.5 g of Trisma base and 87.45 g of NaCl to 850 ml of deionized water. Dissolve the salts and adjust the pH to 7.5 with concentrated HCl. Fill with water to 1000 ml. Store at 4°C.
8. Ten percent SDS (sodium deodecyl sulfate) stock. Add 10 g of powder to 70 ml of distilled water and dissolve by slow stirring. Add water to a final volume of 100 ml and sterilize by filtration through a 0.45 μ filter. Do not autoclave. Store at room temperature.
9. Ten times SSC: 1.5 M NaCl and 0.15 M sodium citrate. Dissolve 87.5 g of NaCl and 44.1 g of sodium citrate in 850 ml of distilled or deionized water. Adjust the pH to 7.5 with 10 N NaOH and add water to 1,000 ml. Store at 4°C.

10. Standard DNA: 1 kb 100 μl Ladder DNA, 700 μl TE buffer, and 200 μl loading dye solution.
11. Stop solution (loading dye): 15 percent Ficoll 400, 5 M urea, 0.1 M sodium EDTA (pH 8.0), 0.01 percent bromophenol blue, and 0.01 percent xylene cyanol. Prepare at least 10 ml of the solution. Dissolve an appropriate amount of Ficoll powder in double-distilled or deionized water by stirring at 40–50°C. Add a stock solution of EDTA, powdered urea, and dyes and aliquot approximately 500 μl into microfuge tubes and store at −20°C.
12. Fifty times TAE (Tris–acetate EDTA) electrophoresis buffer: 2 M Trisma base, 1 M acetic acid, and 50 mM Na_2 EDTA (pH 8.0). Weigh 242 g of Trisma base and add to 800 ml of double-distilled or deionized water. Add 57.1 ml of glacial acetic acid and 100 ml of 0.5 M EDTA stock solution (pH 8.0). Dissolve the powder by continuous stirring for 30 minutes and add water to a final volume of 1 l. Do not autoclave. Store tightly closed at room temperature.
13. Washing solution II: two times SSC and 0.1 percent SDS.
14. Washing solution III: one times SSC and 0.1 percent SDS.

DNA Fingerprinting: Single-locus Analysis

Equipment and supplies

1. D2S44 probe (Promega, no. DK263A or equivalent).
2. MetaPhor™ agarose (FNC BioProducts, no. 501810).

Solutions to prepare

1. Hybridization solution: 0.5 m sodium phosphate (pH 7.2), 0.5 percent (v/v) Tween 20, and 1 percent casein (Hammerstein grade) or blocker casein in PBS (Pierce, no. 37528).
2. Twenty times SSC: 3 M NaCl and 0.3 M sodium citrate. Dissolve 175 g of NaCl and 88.2 g of sodium citrate in 900 ml of distilled or deionized water. Adjust the pH to 7.5 with 10 N NaOH and add water to 1,000 ml. Store at 4°C.
3. Stripping solution: 0.4 N NaOH and 0.1 percent SDS. The solution should be freshly prepared.
4. Ten times TBE (Tris–borate EDTA) electrophoresis buffer: 890 mM Tris base, 890 mM boric acid, and 20 mM EDTA. Dissolve the Tris and boric acid in deionized water and add the appropriate amount of 0.5 M EDTA (pH 8.0). Store at room temperature.
5. Ten times wash buffer I: 0.5 M sodium phosphate (pH 7.2) and 5 percent (v/v) Tween 20.

Out of Africa: Origin of Modern Humans

Equipment and supplies

1. Acetamide (Sigma Co., no. A 0500).
2. *Alu* primers: *Alu* F, CCTTCCACAGTGTATTGTGTC and *Alu* R, TAGAAATGTGTGGGACAGTTC.
3. Capillary tubes (10 μl) (Idaho Technology, no. 1705).
4. Low molecular weight DNA standard (BioMarker Low Bioventure Inc.).
5. Ten times high magnesium PCR buffer mix for an air cycler (Idaho Technology, no. 1781).
6. Ten times low magnesium PCR buffer mix for an air cycler (Idaho Technology, no. 1783).
7. TT Primers: TT F, TAATTGTTGGAGTCGCAAGCTGAAC and TT R, GCCTGAGTGACAGAGTGAGAACC.

Solutions to prepare

1. Fifty percent acetamide.
2. Low DNA standard: 150 μl (BioMarker Low), 70 μl TE buffer, and 40 μl BioTracker tracking dye. Store at −20°C.

DNA Sequencing

Equipment and supplies

1. An AeroMist disposable inhalator–nebulizer (Inhalation Plastic Inc., no. 4207).
2. Ampicillin, sodium salt (Sigma Co., no. A 9518).
3. ATP (100 mM) (Roche Molecular Biochemicals, no. 1 140 965 or equivalent).
4. DNA Sequencing Kit v2 (PE Biosystems, no. 4314417).
5. ElectroMax *Escherichia coli* cells DH 10 B (Life Technologies Inc., no. 318290015).
6. Electroporation cuvettes (0.1 mm gap) (Invitrogen Inc., no. P410-50).
7. Filtration cartridges (Edge Biosystem Inc., no. 42453 or equivalent).
8. Oak Ridge polypropylene 50 ml centrifuge tubes with caps (e.g. Nalgene® no. 21009).
9. PLG (Phase Lock Gel) I tubes (Eppendorf, no. 0032007953).
10. PCI (Phenol : CIA) mixture (Ambion, no. 9732).
11. Polypropylene culture tubes (Falcon, no. 2059 or equivalent). The

"Falcon 2059" tube of Becton Dickson Co. is the standard for transformation experiments. Other equivalent brands are acceptable, but batches of tubes are occasionally contaminated with surfactants that inhibit transformation

12. pUC18 SmaI/BAP (Amersham Pharmacia Biotech, no. 27-1860 01).
13. Rapid DNA Ligation Kit (Roche Molecular Biochemicals, no. 1 635 379 or equivalent).
14. T4 DNA polymerase (NEB, no. 203S).
15. T4 polynucleotide kinase (NEB, no. 201S).
16. Transformation apparatus E. coli pulser (Bio-Rad, no. 3 165-2101 or equivalent).
17. Universal M13/pUC sequencing primer (NEB, no. 3 1211 or equivalent).

Solutions to prepare

1. Ammonium acetate (7.5 M). Dissolve 57.8 g of ammonium acetate in 60 ml of double-distilled or deionized water. Stir until the salt is fully dissolved. Do not heat to facilitate dissolving. Fill up to 100 ml and sterilize by filtration. Store tightly closed at 4°C. The solution can be stored for one to two months under these conditions. Long-term storage is possible at −70°C.
2. One thousand times ampicillin: 500 mg ampicillin and distilled or deionized water. Add 500 mg of ampicillin to 5 ml of distilled water. Sterilize by filtration and store in small aliquots at −20°C.
3. ATP (0.5 mM). Dilute 5 mM ATP solution ten times. Add 2 µl of 5 mM ATP stock solution to 18 µl of sterile water. Prepare the solution freshly. Do not store.
4. ATP (5 mM). Prepare 100 µl of solution. Add 5 µl of stock ATP solution to 95 µl of sterile 5 mM Tris–HCl (pH 7.5). Store at −20°C.
5. Seventy percent ethanol. Add 25 ml of double-distilled or deionized water to 70 ml of 95 percent ethanol. Never use 100 percent ethanol because it contains an additive that can inhibit the activities of some enzymes. Store in a −20°C freezer.
6. IPTG (25 mg ml^{-1}): 2.5 percent IPTG. Dissolve 250 mg of IPTG in sterilized water. Store at −20°C.
7. LB agar amp plates: 1 percent Bacto Tryptone, 0.5 percent yeast extract, 0.5 percent NaCl, 1.5 percent Difco agar, and 100 µg ml^{-1} ampicillin. Add the first three ingredients to 1 l of distilled water in a 2 l Erlenmeyer flask. Stir to dissolve all the ingredients completely. Adjust the pH to 7.5 with 1 N NaOH. This will take approximately 4 ml of 1 N NaOH. Add the Difco agar and sterilize by autoclaving for 20 minutes. Cool the medium to 60–65°C and add 1 ml of ampicillin stock solution. Mix by swirling the flask and pour the plates. This will make 25–30 plates. The plates can be stored for two to three weeks at 4°C.

8. Ten times phosphate stock solution: 0.72 M KH_2PO_4 and 0.17 M K_2HPO_4. Dissolve in water and autoclave for 20 minutes. Store at 4°C.
9. Solution II (plasmid preparation): 0.2 N NaOH and 10 percent SDS. Prepare freshly before use.
10. Terrific broth medium (TB): 1.2 percent Bacto Tryptone, 2.4 percent yeast extract, 0.4 percent glycerol, and ten times phosphate stock solution. Mix the first three ingredients in 900 ml of deionized water and autoclave for 20 minutes to cool to room temperature and add 100 ml of phosphate stock solution. Store at 4°C.
11. X-gal (20 mg ml^{-1}): 2 percent X-gal and DMSO. Dissolve 200 mg of X-gal in 10 ml of DMSO. Store in the dark at –20°C. DMSO is used instead of the commonly used DMF (dimethylformamide) for X-gal preparation because DMF is very toxic.

Determination of Human Telomere Length

Equipment and supplies

1. Wizard Genomic Purification Kit (Promega, no. TM050).
2. *Hinf*I restriction endonuclease (NEB, no. 155S or equivalent).
3. *Rsa*I restriction endonuclease (NEB, no. 167S or equivalent).
4. Telo TAGGG Telomere Length Assay (Roche Molecular Biochemicals, no. 2 209 136).
5. TurboBlotters (one per group) (Midwest Scientific Co., no. 10-439-012).

RT-PCR of Human Genes

Equipment and supplies

1. RNaseZap^{TM} solution (Ambion, no. 9780 or equivalent).
2. RNAwiz^{TM} solution (Ambion, no. 9736 or equivalent).
3. RNAsecure^{TM} resuspenstion solution (Ambion, no. 7010 or equivalent).
4. Formaldehyde load dye (Ambion, no. 3 8552).
5. Titan^{TM} One Tube RT-PCR Kit (Roche Molecular Biochemicals, no. 1 939 823).
6. Human β-actin primers: forward, CCAAGGCCAACCGCGAGAA-GATGAC and reverse, AGGGTACATGGTGGTGCCGCCAGAC.

Index

Page numbers in *italics* refer to Figures; those in **bold** to Tables.

ABI 3700 sequencer, 134–6
affinity matrix application, 131
agarose, definition of, 30
agarose gel electrophoresis
   Alu amplification, 90–1
   buffers, 31–2
   in fingerprinting, 29–35, *104*
   in plasmid preparation for DNA
      sequencing, 133–4, *135*
   principle of, 29–31
   RT–PCR of human genes, 204–7, *206*,
      211
   telomere length determination and,
      177–9
alignment sequences, 143–7
   with BLAST, 158–9
alkaline lysis, in plasmid preparation for
   DNA sequencing, 130–1
alkaline transfer of DNA, 37–8
allele frequencies, single-locus DNA
   fingerprinting, 70
Alu amplification, reaction stock mixture
   for, **85**
Alu primers, agarose gel electrophoresis, *92*
Alu SINE element, 159
Alu STR haplotypes in world populations,
   82
ammonium acetate solution
   DNA precipitation, 8
   in plasmid preparation for DNA
      sequencing, 131
amplification
   rapid cycle, 84
   STR primers, *92*
ATP concentration, in DNA sequencing,
   118
aurintricarboxylic acid (ATA), 199

autoradiography, DNA probe preparation
   and, 43–4
avian myoblastosis virus (AMV), 211

bacteria, transformation by electroporation,
   121–9
band-sharing coefficients, DNA
   fingerprinting, 59
base content analysis, single-sequence
   analysis and, 162
Basic Local Alignment Search Tool
   (BLAST), 139, 147–50
   sequence alignment with, 158–9
   versus FASTA, 150–1
beta-lactamase gene, DNA sequencing,
   113
bins, single-locus DNA fingerprinting, 71
biotin, DNA probe preparation, 44
BLAST *see* Basic Local Alignment Search
   Tool
blunt-end cloning, 115
blunt-end ligation
   in DNA sequencing, 118
   repair of ends of sheared DNA, *107*
boiling lysis, in plasmid preparation for
   DNA sequencing, 131
buffers, in electrophoresis, 31–2

capillary nebulization channel size, 102
capillary transfer, 37–8
CD4 STR alleles, 80
cheek cells
   absorption spectrum of DNA, *18*
   breakage, 3
   collection of, 13, 171, 202
chemiluminescent detection, DNA probe
   preparation, 44

Index

chloroform:isoamyl alcohol (CIA)
  Marmur method, 5
  repair of ends of sheared DNA, 106–7
  safety precautions, 11
chromatograms, editing, 155–6
chromosome 2, DNA single-locus analysis and, 62
chromosomes
  position of query sequence, 159–60
  telomeric structure on, *167*
cloning
  blunt-end, 115
  preparation of plasmid for, 114–15
  principle of, 111–12
  vectors, 112–14
colormetric detection, DNA probe preparation, 44
complementary strands, DNA hybridization, 46
computer analysis of sequencing data, 139–63
  base content analysis, 162
  BLAST, 147–50, 150–1, 158–9
  chromosome position of query sequence, 159–60
  databases accessible through Internet, **141**
  dot matrix analysis, 152–4, *153*, 162–3
  dynamic programming decision matrix, *145*
  EMBL sequence file format, *143*
  FASTA, *143*, 150–1
  formats, 140–3
  GenBank DNA sequence file format, *142*
  global alignment principles, *146*
  local alignment principles, *146*
  restriction enzyme site analysis, 162
  sequence alignments, 143–7, 158–9
  single-sequence analysis, 151–2, 161–3
  TRF length analysis, 193–4
concatameric ligation, DNA sequencing, 116
cycle sequencing, 134–6

D2S44 alleles, single-locus DNA fingerprinting, 74
  in Simpson trial, 74–5
data analysis, 59–61, 69–71
database option, in BLAST, 149
databases, sequence formats and, 140–3
DDBJ, 140–2
deproteinization, 3–6, 170–3

enzymes, 6
Kirby method, 4
Marmur method, 4
organic solvents, 3–5
detection
  buffer, single-locus DNA fingerprinting, 67–8
  chemiluminescent, 44
  colorimetric, 44
  membrane preparation for, 56–7
  signal procedure, 58–9
dideoxy terminators, removal in DNA sequencing, 137
diethyl pyrocarbonate (DEPC), 199–200
DIG, *see* digoxigenin
Dig Easy Hyb solution, 49
digestion, restriction enzyme, 25–8
digoxigenin (DIG), 43, 164
  DNA fingerprinting, 51
  preparation of probe, **51**, 51–3
dissociation, DNA hybridization and, 46
DNA
  absorption spectrum of, *18*
  alkaline transfer, 37–8
  amplification, 84, *92*
  cell breakage, 3
  cell composition, **2**
  cheek cell collection and, 11–19
  circular DNA, nebulization, 103
  cloning, principle of, 111–12
  concentration, 7–9, 17–18, 174
  deproteinization and, 3–6
  end-repair reaction, **108**
  fingerprinting, *see* fingerprinting
  hybridization, *see* hybridization
  isolation procedure, 2, 170–3
  precipitation of, 7–9, 16–17
  preparation of, 1–19
  purification of, 9–19, 171–3, 174
  quantity and, 9–10
  replication, at ends of linear molecules, *168*
  RNA removal and, 6–7
  sequencing, *see* sequencing
  shearing by nebulization, 100–10
DNA polymerase, random priming and, 45
DNase contamination, 12
DNASIS program, single-sequence analysis and, 162
dot matrix analysis, 152–4
  principle of, *153*
  single-sequence analysis and, 162–3

dynamic programming decision matrix, 144–5
    graph of, *145*

EDTA, heavy metals and, 12–13
electrophoresis, *see* agarose gel electrophoresis
electroporation, transformation of bacteria by, 121–9
EMBL, *see* European Molecular Biology Laboratory
end-repair reaction, **108**
end-replication problem, telomere length and, 167
ENTREZ, 140
enzymatic procedures, RNA removal and, 6–7
enzyme digestion, restriction, 25–8, 174–7
enzyme–alkaline phosphatase, 62
enzymes
    deproteinization, 6
    restriction digestion, 25–8, 174–7
Escherichia coli, nick translation, 44–5
EST, *see* Expressed Sequence Tag
ethanol, DNA precipitation, 7–9
ethidium bromide, 13
eucaryotic chromosomes, telomeres and, 165
European Molecular Biology Laboratory (EMBL), 140–3
    sequence file format, *143*
expect value option in BLAST, 149–50
Expressed Sequence Tag (EST), 139
expressed sequences, search for, in BLAST, 159

FASTA, 147, 150
    sequence file format, 142, 143, *143*
    versus BLAST, 150–1
filter option, in BLAST, 148–9
fingerprinting
    agarose gel electrophoresis, 29–35
    alkaline transfer in, 37
    digoxigenin-labelled probe, **51**
    endonuclease, 25–8
    hybridization, 43–54
    membrane preparation for detection, 56–7
    mismatched hybrid removal and, 56–7
    multi-locus analysis, 20–61, *21*, *60*, 217–18
    probe preparation, 43–54

restriction enzyme digestion in, 25–8
signal detection procedure, 58–9
single-locus analysis, 62–75, *64*, 218–19
Southern blotting, 37–41, *40*
fixed bins, single-locus DNA fingerprinting, 71
floating bins, single-locus DNA fingerprinting, 71
fluorescent tags, DNA probe preparation and, 44
forensic profiling, single-locus DNA fingerprinting, 65, 71
formamide concentration, hybridization reaction solution, 49

gel electrophoresis, *see* electrophoresis
gel size, DNA fingerprinting and, 32
gel staining, DNA fingerprinting and, 33
gels, photographing, 33–4
GenBank database, 139
    DNA sequence file format, 140–2, *142*
global alignments, principle of, *146*
guanidinium hot-phenol method, 200
guanidinium isothiocynate, 201

hapten-like molecules, DNA probe preparation and, 44
haptens, DNA probe preparation and, 44
high-salt lithium chloride method, 200
high-scoring segment pairs (HSPs), 148
high stringency washes, washing reaction and, 50
Homo erectus, 77
Homo sapiens, 77
human genomic DNA, *see* DNA
human sequencing library plate, *129*
humans, modern, origin of, 76–93
hybridization, 46, 185–6, 187–9
    DNA probe preparation, 43–54
    pre-hybridization, 53–4
    reaction solution, 49–50
    removal of mismatched hybrids, 56
    temperature, 47–8
    time in, 48–9
hydrodynamic shearing, avoidance of, 11–12

Intelligenetic sequence entry format, 143
Internet, sequence databases accessible through, **141**
isoamyl alcohol, Marmur method, 5

isolation procedure, schematic outline of, 2
isopropanol, DNA precipitation, 7–9

Kirby method of deproteinization, 4
KTUP values, 150

laminar flow, 101
ligation reactions in DNA sequencing, 111–21, **119**
linear molecules, DNA replication at end of, *168*
linkage disequilibrium, definition of, 78
living cells, composition of, **2**
loading dye solutions, DNA fingerprinting and, 32–3
local alignment analysis programs, 139–63
  principle of, *146*
lysis by boiling, in plasmid preparation for DNA sequencing, 131

macaloid, RT–PCR of human genes, 199
Marmur method of deproteinization, 4
  chloroform:isoamyl alcohol and, 5
master mix 1 (MX1), RT–PCR of human genes, **213**
master mix 2 (MX2), RT–PCR of human genes, **213**
master reaction mixture, in DNA sequencing, **119**
match windows, single-locus DNA fingerprinting, 72, **74**
maximal-scoring segment pairs (MSPs), 148
melting, DNA hybridization and, 46, 47
membrane detection, and hybridization, 187–8
membrane preparation, for detection, 56–7
membrane stripping
  DNA single-locus analysis and, 62
  single-locus DNA fingerprinting, 66, 73
MetaPhor™ agarose, 89–90
micro-satellites, in multi-locus DNA fingerprinting, 22
mini-satellites
  in multi-locus DNA fingerprinting, 22
  in single-locus DNA fingerprinting, 63
mitochondrial Eve data, and modern human origin, 78
modern human origin
  Alu amplification, 90–1
  Alu STR haplotypes in world populations, **82**

Alu(–)STRs in world populations, **81**
agarose gel electrophoresis of amplification, *92*
amplification of human genomic DNA using STR primers, *92*
data analysis, 91
experiment schematic, 77
high resolution agarose electrophoresis, 89–90
multiple-origin model, 77
polymorphic markers at CD4 locus on chromosome 12, *80*
rapid cycle DNA amplification, 84
reaction stock mixture, **85, 86**
running PCRs, 85–7, **86**
STR primers, *92*
TT amplification, 89–90
multi-locus analysis, DNA fingerprinting, 20–61, *21*, 217–18
  principle of, *23*
multiple cloning site, DNA cloning, 113

NBRF sequence entry format, 143
nebulization shearing of DNA, 100–6
Needleman–Wunsch algorithm, 145
nick translation, 44–5
NIH Image, TRF length analysis and, 191, 193, 194
non-radioactive reporters, DNA probe preparation and, 44
nr database search, BLAST and, 158–9
nucleation reaction, DNA hybridization and, 46

Oligo(dT)12–18 primer, RT–PCR of human genes, 211
organic solvents, deproteinization and, 3–5
out-of-Africa replacement (OAR) hypothesis, 77, 78

pairwise alignment analysis, 153
PCI reagent, repair of ends of sheared DNA, 106–7
PCR reaction kinetics, 81–4
phenol
  in deproteinization, 4
  repair of ends of sheared DNA, 106–7
  RT–PCR of human genes, 201
  safety precautions, 11
phosphodiester bonds, heavy metals and, 12–13
photo-oxidation, 13

phylogenetic analysis, 77
plasmid preparation
  for cloning, 114–15
  for DNA sequencing, 130–4
plasmid sequences, removal of, 154
polymerase chain reaction (PCR)
  amplification, in ABI 3700 sequencing, 134–6
  multi-locus DNA fingerprinting, 24
  probe preparation, 45–6
  rapid cycle DNA amplification, 84
  running, 85–7, **86**
  see also RT-PCR
PCR STR reaction (RSMT), modern human origin, 85–6
polymorphic loci, definition of, 78
polymorphism, single-locus DNA fingerprinting and, 63
polynucleotide kinase removal, 108
positive selection marker, DNA sequencing, 113
pre-hybridization, 53–4, 186
  single-locus DNA fingerprinting, 66
precipitation of DNA, 16–17
probes, 45–6
  preparation, and hybridization, 43–4
product rule, single-locus DNA fingerprinting, 71
programming, dynamic, 144–5, *145*
protein denaturation agents, 199
protein removal, *see* deproteinization
protein RNase inhibitors, 199
protein, salting out, 170–3
purification, 13–15, 16–19
  determination of DNA purity and quantity, 9–10, 17–18, 171–3, 174
  of plasmids for DNA sequencing, 130
  of total RNA, 198–203

quantity determination of DNA, 9–10
query sequence, chromosome position of, 159–60

radioactive reporters, DNA probe preparation and, 43–4
random fragment sequencing, 95
random hexanucleotide primers, RT-PCR of human genes, 212
random priming, 45
random strategies (shotgun strategies), 95, 97
rapid cycle DNA amplification, 84

re-annealing reaction, DNA hybridization, 46
re-association, DNA hybridization, 46
reaction stock mixture
  for *Alu* amplification (RSMA), **85**
  in DNA sequencing, **119**, **136**
  hybridization and, 49–50
  STR (RSMT), modern human origin, **86**
READSEQ program, 143
recirculization, DNA sequencing and, 116
removal of mismatched hybrid, 56
renaturation, DNA hybridization and, 46
repair reaction, repair of ends of sheared DNA, *107*
replicative history, telomere length, 168
replicon, ligation to sequencing vector, 112, 113
reporter enzyme–alkaline phosphatase, 62
reporters
  non-radioactive, 44
  radioactive, 43–4
restriction endonuclease, in DNA fingerprinting, 25–8
restriction enzyme digestion, 25–8, **27**, 174–7
restriction enzyme site analysis, 162
restriction reactions, preparation of, **176**
resuspension of cells, in plasmid preparation for DNA sequencing, 131
reverse transcriptase reaction, RT–PCR of human genes, 210–12
RNA, 196–214
  agarose gel electrophoresis, 204–7
  isolation, 200
  purification of, 198–203
  removal from DNA preparations, 6–7
RNaseOff solution, RT–PCR of human genes, 200
RNases, elimination of, 199–200
RNaseZap^{TM}, RT–PCR of human genes, 200
RNasin, RT–PCR of human genes, 199
RNAwiz^{TM}, RT–PCR of human genes, 201
RT–PCR, 96–214
  analysis of results, 214
  human cheek cell collection, 202
  master mix1 (MX1), **213**
  master mix2 (MX2), **213**
  procedures schematic, *197*
  reverse transcriptases, 210–12
  RNA agarose gel electrophoresis, 204–7, *206*, *211*

RNA isolation methods, 200
RNA purification, 202–3
RNases elimination, 199–200
running of, 209–14

salt concentration, hybridization reaction solution, 49
salting out, of protein, 170–3
satellites, in multi-locus DNA fingerprinting, 22
Sequencher
  chromatogram editing, 155
  single-sequence analysis and, 161–3
sequencing, 95–137, 220–2
  ABI 3700 sequencer, 134–6
  agarose gel electrophoresis in, *104*, 133–4, *135*
  alignments, 143–7, 158–9
  bacteria transformation by electroporation, 121–9
  blunt-end ligation schematic, *107*
  cloning, 111–15
  coverage, 99
  data, computer analysis of, 139–63
  databases, Internet access and, **141**
  dideoxy terminator removal, 137
  electroporation in, 121–9
  end-repair reaction in, **108**
  file format, 140–3
  human sequencing library plate, *129*
  library, 97, 100–1, *129*
  ligation reaction and, 115–18, **119**
  ligation to sequencing vector, 111–21
  master reaction mixture, **119**
  methods, 97–8
  mini-gel electrophoresis, 133–4
  nebulization shearing in, 100–6
  plasmid preparation, 114–15, 130–4, *135*
  plasmid removal, 154
  procedures schematic, *96*
  reaction mix, **136**
  repair reaction schematic, *107*
  shearing DNA, 100–10
  strategies, 97, 98–9
  transfer, TurboBlotter and, *183*
  transformation of bacteria by electroporation, 121–9
  vector, ligation to, 111–21
short interspersed nuclear elements (SINEs), 139
short tandem repeats (STRs)

amplification of human genomic DNA using, *92*
modern human origin and, 76, 79
multi-locus DNA fingerprinting, 24
signal detection
  DNA fingerprinting and, 58–9, 67–8
  Simpson Trial D2S44 probe, 74–5
single-locus DNA fingerprinting, 22, 62–75, 218–19
  allelic frequencies of D2S44 VNTR locus, **70**
  autogram of, *73*
  data analysis and, 69–71
  membrane stripping, 66
  outline schematic of, *63*
  pre-hybridization, 66–7
  principle of, *64*
  signal detection, 67–8
  Simpson Trial D2S44 probe, 74–5
  washing reaction, 67
single-sequence analysis, 151–2, 161–3
sliding window filter, dot matrix analysis and, 153
Smith–Waterman algorithm, 145, 147
sodium acetate, DNA precipitation, 8
sodium chloride, DNA precipitation, 8
solution II, in plasmid preparation for DNA sequencing, 131, 132
Southern blotting, 37–41
  preparation, *40*
  single-locus DNA fingerprinting, 63
  telomere length determination, 180–4
specific oligonucleotide primers, 212
strand separation, DNA hybridization and, 46
STRs, *see* short tandem repeats
subclone coverage, DNA sequencing and, 99
submarine gels, DNA fingerprinting and, 32
supercoiled DNA, nebulization and, 103

T4 DNA polymerase, in DNA sequencing, 107
telomere length determination, 164–94, 222
  activity, schematic of, *169*
  agarose gel electrophoresis, 177–9
  average length of TRF region, *192*, **193**
  chromosome end and, *167*
  collecting human cheek cells, 171
  concentration of DNA, 174

deproteinization, 170–3
DNA replication at ends of linear
  molecules, *168*
genomic DNA isolation, 170–3
hybridization and, 185–9
pre-hybridization and, 186
purification, 171–3, 174
restriction enzyme digestion, 174–7
restriction reaction preparation, **176**
telomerase activity schematic, *169*
outline of procedures schematic, *166*
Southern transfer, 180–4
TRF length analysis and, 190–4
TurboBlotter and, *183*
Telomere Reverse Transcriptase (TERT),
  telomere length and, 167
terminal restriction fragment (TRF), 164,
  190–4
determination of average length of
  region, *192*
calculation of, 192–4
time, DNA hybridization and, 48–9
transformation, of bacteria by
  electroporation, 121–9
translation, nick, 44–5
TRF, *see* terminal restriction fragment
TRI-Reagent™, RT–PCR of human genes,
  200

TT amplification, high resolution agarose
  electrophoresis, 89–90
TurboBlotting, telomere length
  determination and, 180, 181, *183*

uncertainty data, single locus DNA
  fingerprinting, **74**

vanadyl–ribonucleoside complexes, 199
variable number tandem repeat (VNTR)
  in multi-locus DNA fingerprinting, 22
  in single-locus DNA fingerprinting, 63
vector sequences, removing, in dot matrix
  analysis, 156
vortexing, digoxigenin label concentration,
  51

washing reaction
  DNA fingerprinting and, 50, 67
  hybridization, 187
Wizard DNA purification kit, 170–3
word, definition of, 150
word size option, in BLAST, 148
world populations
  Alu STR haplotypes in, **82**
  Alu(–)STRs values in, **81**

zippering reaction, DNA hybridization, 46

Lightning Source UK Ltd.
Milton Keynes UK
UKOW011138180713

213928UK00003B/75/P